Theory
and Reality

Theory
and Reality

*An Introduction to the Philosophy
of Science* | Second Edition

Peter Godfrey-Smith

The University of Chicago Press • *Chicago and London*

The University of Chicago Press, Chicago 60637
The University of Chicago Press, Ltd., London
© 2021 by Peter Godfrey-Smith
Published 2021
Printed in the United States of America

30 29 28 27 26 25 24 23 22 21   1 2 3 4 5

ISBN-13: 978-0-226-61865-4 (paper)
ISBN-13: 978-0-226-77113-7 (e-book)
DOI: https://doi.org/10.7208/chicago/9780226771137.001.0001

Library of Congress Cataloging-in-Publication Data

Names: Godfrey-Smith, Peter, author.
Title: Theory and reality : an introduction to the philosophy of science /
    Peter Godfrey-Smith.
Description: Second edition. | Chicago : University of Chicago Press,
    2021. | Includes bibliographical references and index.
Identifiers: LCCN 2020053632 | ISBN 9780226618654 (paperback) |
    ISBN 9780226771137 (e-book)
Subjects: LCSH: Science—Philosophy.
Classification: LCC Q175.G596 2021 | DDC 501—dc23
LC record available at https://lccn.loc.gov/2020053632

♾ This paper meets the requirements of ANSI/NISO Z39.48-1992
(Permanence of Paper).

For my parents

Contents

Preface

This book is based on undergraduate lectures given at Stanford University, Harvard University, and the University of Sydney over a period of nearly thirty years. The book is a distillation of lectures, but not only of lectures. It also bears the influence of innumerable comments, questions, and papers by students over that time, together with remarks made by colleagues and friends.

The book is written primarily for students, but it is intended to be accessible to a wide audience. I assume no background knowledge in philosophy at all in the reader. My primary aim is to introduce some of the main themes in the philosophy of science, while simultaneously telling an accessible story about how the field has developed from early in the twentieth century to the present day. In telling this story I attend quite closely to the connections between philosophy and other disciplines, and track the changing intellectual climate in which theories about science have been offered. I have also tried, in some places, to capture some of the personalities of the protagonists and the atmosphere of the debates.

Another aim of the book is the outline and defense of a particular point of view. I have concentrated that discussion mostly in the final third of the book. The philosophy of science remains in a state of considerable ferment. That poses a choice for the author of a book like this; one can either abstract away from the disorder and uncertainty and lay down one particular vision, or one can use the disputes to tell a story about the field—how did we get to where we are now? I have mostly chosen the latter approach. So much of the book is organized chronologically, especially until chapter 9, after which the chapters move more from topic to topic.

Each chapter ends in a section called "Further Reading and Notes." These contain not only references for additional reading, but extra acknowledgments, comments, and paths for exploration. The "Further Reading" itself tends to include quite a lot of primary material, including some difficult works and works intended to give the flavor of recent discussion. The glossary at the end of the book, in contrast, is intended to be very elementary, a tool for those coming to the book with little or no background in the area. As well as a standard bibliography, the end of the book has a list of websites.

This is the second edition of *Theory and Reality*. The first was published in 2003. The project of the second edition was to update an existing book, not to write a new one. The narrative structure of the first edition has been kept in place, some passages are nearly untouched, and additions and changes have been mostly aimed at getting the book up to date. In order to make complicated material accessible, I have also brought the writing a little closer to the style of a lecture in some places. Other changes reflect changes of mind since 2003, and chapter 13 is entirely new.

For help writing the first edition, I remain grateful to Fiona Cowie, Michael Devitt, Stephen Downes, Richard Francis, Michael Friedman, Lori Gruen, Tania Lombrozo, Denis Philips, J. D. Trout, Allen Wood, Rega Wood, and Susan Abrams, my old editor at the University of Chicago Press. Detailed comments on entire drafts of the first edition were written by Karen Bennett, Kim Sterelny, and Michael Weisberg. That first edition also benefited from the insight, good sense, deft touch, and unique perspective of David Hull, the first editor of the *Science and Its Conceptual Foundations* series. In writing the update, I am grateful for the help of Lindell Bromham, Jordi Cat, Jim Joyce, Adam Hochman, Maureen O'Malley, Tibor Molnar, Bence Nanay, and Jane Sheldon. Joel Velasco wrote extremely helpful comments on the near-final draft, and Amelia Scott, in the course of organizing the references, contributed a number of insights. Amelia also drew the new figure on page 112. Finally, I would like to thank my supportive and patient editor at Chicago, Karen Merikangas Darling, and my agent, Sarah Chalfant.

Chapter 1

Introduction

1.1 *Setting Out*

This book is an introduction to a collection of ongoing debates about the nature of science—how it works, what it achieves, and what (if anything) makes science different from other ways of investigating the world. Most of the ideas we will examine fall into the field called "philosophy of science," but we will also spend a good deal of time looking at ideas developed by historians, sociologists, psychologists, and others.

The book is organized mostly as a historical narrative that covers a little over one hundred years, from the early twentieth century to early twenty-first. Ideas will be discussed in roughly the order in which they appeared, especially in the first half of the book. Why is it best to start with older ideas and work through to the present? One reason is that the historical development of ideas about the nature and workings of science is itself an interesting topic. I also think that this is the best way to come to understand many of the debates going on now. The main narrative of the book begins in the early years of the previous century. That might seem to be starting a long way back. But I think that if we start there, the story does make sense. If we start later, you are likely to find yourself wondering: why are people setting the issues up like *that*? A good way to understand the maze of options and opinions in the field is to trace the path that brought us to the state in which we find ourselves now.

The book will tell a historical story and use that to understand debates about science, but my aim is not just to introduce all the different views people have had. I will often take sides as we go along, trying to indicate which developments were wrong turns and which were closer to the right track.

Philosophy is an attempt to ask and answer some very basic questions about the universe and our place within it. These questions can sometimes seem far removed from practical concerns. But the debates covered in this book are not of that kind. Though these debates are connected to the most abstract questions about thought, knowledge, language, and reality, they have turned out to have an importance that extends well beyond philosophy. They have made a difference to many other academic fields, and some of the debates have reverberated much further, affecting discussions of education, medicine, and the proper place of science in democratic societies.

In fact, throughout much of the period covered in this book, all the fields concerned with the nature of science went on something of a roller-coaster ride. Especially in the later twentieth century, some people thought that work in the history, philosophy, and sociology of science had shown that science does not deserve the dominating role it has acquired in Western cultures. They thought that a set of convenient myths about the trustworthiness and superiority of mainstream science had been thoroughly undermined. Others disagreed, of course, and the resulting debates swirled across the intellectual scene, frequently entering political discussion as well. From time to time, scientific work itself was affected, especially in the social sciences. These debates came to be known as the "science wars," a phrase that conveys a sense of how heated things became.

The science wars eventually cooled down, but there is still a great deal of disagreement about even the most basic questions concerning the nature and status of scientific knowledge. These disagreements often do not have much influence on the day-to-day practice of science, but sometimes they do. And they have great importance for discussions of human knowledge, cultural change, intellectual freedom, and our overall place in the universe. This book aims to introduce you to this remarkable series of debates, and to give you an understanding of the present situation.

1.2 *The Scope of the Investigation*

If we want to understand how science works, it seems that the first thing we need to do is work out what exactly we are trying to explain. Where does science begin and end? Which kinds of activity count as "science"?

Unfortunately, this is not something we can settle in advance. There is a lot of disagreement about what counts as science, and these disagreements are connected to all the other issues discussed in this book.

There is consensus about some central cases. People often think of physics as the purest example of science. Certainly physics has had a heroic history and a central role in the development of modern science as a whole. Biology, on the other hand, is probably the science that has developed most rapidly and impressively during recent years.

These seem to be central examples of science, though even here we encounter hints of controversy. A few have suggested that theoretical physics is becoming less scientific than it used to be, as it is evolving into an esoteric, mathematical model–building exercise that has little contact with the real world (Horgan 1996). And biology has recently acquired connections with business and industry that make it, in the eyes of some, a less exemplary science than it once was. Still, examples such as these give us a natural starting point. The work done by physicists and biologists when they test hypotheses is science. And playing a violin, no matter how well one plays, is not doing science. But in the area between these clear cases, disagreement reigns.

At one time the classification of economics and psychology as sciences was controversial. Those fields have now settled into a scientific status. (Economics retains an amusing qualifier; it is sometimes called "the dismal science," a phrase due to Thomas Carlyle.) There is still a much-debated border region, however, and this includes areas like anthropology and sociology. At Stanford University, during the time I was working there on the first edition of this book, a debate over the boundaries of science was part of a process in which the Department of Anthropology split into two separate departments. Is anthropology, the general study of humankind, a fully scientific discipline that should be closely linked to biology, or is it a more "interpretive" enterprise that should be more closely connected to the humanities? After nine years apart, the two Stanford departments reunited. But anthropology is still one of the most fought-over grounds. In 2010 the American Anthropological Association released a new version of their summary of what the field is, dropping the word "science" from the central place it had in earlier statements. The result was uproar and controversy, with an eventual compromise.

The existence of this gray area should not be surprising, because the word "science" is a loaded and rhetorically powerful one. People often find it a useful tactic to describe work in a borderline area as "scientific" or as "unscientific." Some will call a field scientific to suggest that it uses rigorous methods and delivers results we should trust. Less often, but occasionally, a person might describe an investigation as scientific in order to say something negative about it—to suggest that it is dehumanizing, perhaps. (The term "scientistic" is more often used when a negative impression is to be conveyed.) Because the words "science" and "scientific" have these rhetorical uses, we should not be surprised that people constantly argue back and forth about which kinds of intellectual work count as science.

The history of the term "science" is also relevant here. Our familiar ways of using the words "science" and "scientist" developed quite recently. The word "science" is derived from the Latin word "scientia." This translates roughly as knowledge, but it referred particularly to the results of logical demonstrations that reveal general and necessary truths. Scientia could be gained in various fields, but the kind of proof involved was what we would now mostly associate with mathematics and geometry. Around the seventeenth century, when what we now call the Scientific Revolution was taking place, many fields we would now describe as science were usually called "natural philosophy" (physics, astronomy, and other inquiries into the causes of things) or "natural history" (botany, zoology, and other descriptions of the contents of the world). Over time, the term "science" came to be used for work with closer links to observation and experiment, and the association between science and an ideal of conclusive proof receded. Scientific knowledge, in this newer conception, can be reliable without being provable and certain. The accompanying term "scientist" was coined by William Whewell in the nineteenth century. Given the rhetorical load carried by the word "science," we should not expect to be able to lay down, here in chapter 1, an agreed-on list of what is included in science and what is not. At least for now, we will let the gray area remain gray.

A further complication comes from the fact that philosophical theories (and others) often differ in how broadly they conceive of science. Some writers use terms like "science" or "scientific" for any work that assesses ideas and solves problems in a way guided by observational evidence. In this broad sense, science is a basic human activity found in all cultures. There are also views that construe science more narrowly, seeing it as a cultural phenomenon that is localized in space and time. For views of this kind, it was only the Scientific Revolution of the seventeenth century in Europe that gave us science in the full sense. Before that, we find "roots" or precursors of science—work that was often very impressive but different from science itself.

To set things up this second way is to see science as unlike, in many ways, the kinds of investigation and knowledge that routinely go along with farming, architecture, and other kinds of technology. A view like this need not claim that people in nonscientific cultures must be ignorant or irrational; the idea is that in order to understand science, we need to distinguish it from other kinds of investigation of the world. We need to work out why this approach to knowledge, developed by a small group of Europeans, turned out to have such dramatic consequences for humanity.

As we move from theory to theory in this book, we will find some people construing science broadly and others more narrowly. This does not stop us from outlining, here in the first chapter, what kind of understanding we would eventually like to have. However we choose to use the word "science," in the end we should try to develop both:

1.   a general understanding of how humans gain knowledge of the world around them, *and*
2.   an understanding of what makes the work descended from the Scientific Revolution different—if it really is different—from other kinds of investigation of the world.

We will move back and forth between these two kinds of questions throughout the book.

Before leaving this topic, there is one other possibility that should be mentioned. How confident should we be that all the work we call "science," even in the narrower sense described above, has much in common? One of the hazards of philosophy is the temptation to come up with theories that are too broad and sweeping. "Theories of science" need to be scrutinized with this problem in mind.

1.3 *What Kind of Theory?*

This book is an introduction to the philosophy of science. But much of the book focuses on one set of issues in that large field. These questions are about knowledge, evidence, and rationality. For example, how is it possible for observations to provide evidence for a scientific theory? Can we ever be confident that we are learning how the world really works? Is there a reason to prefer simple theories over more complex ones? Questions like these fall into the part of philosophy known as *epistemology*. Philosophy of science also overlaps with other parts of philosophy— philosophy of language, philosophy of mathematics, philosophy of mind, and *metaphysics*, a part of philosophy that is especially controversial and that deals with the most general questions about the nature of reality itself.

Questions about rationality and evidence—the ones that will often be central in this book—are connected to questions about the authority of science. Do we have reason to rely on scientific work when we have to make decisions about what to do to solve a practical problem—an environmental problem, for example? Those problems about the authority of science are especially pressing, but puzzles about the authority of science arise even before we raise questions about the use of science in policy decisions. I'll introduce what I mean with an example. Nearly all human cells (and the cells of other animals, plants, and many unicellular organisms) contain mitochondria. These are little structures, with their own membranes around them, that contribute energy to the cell. They are often described as "the cell's powerhouses" or in similar terms. When I was a student, especially as a result of the work of Lynn Margulis, there was a lot of discussion of the surprising hypothesis that these parts of our cells are descended from free-living bacteria. The idea is that some bacterial cells (or just one) were swallowed in the distant past by another unicellular organism, and our mitochondria-carrying cells today are all descended from that arrangement. This was initially a very speculative possibility that had first been raised in the late nineteenth and early twentieth centuries (before people had a clear picture of what mitochondria were like) by Richard Altmann and others. Margulis defended the idea with new evidence in a 1967 article. After much controversy, during the mid- and late 1980s the idea finally became widely accepted, and it is now in the textbooks.

It may be in the textbooks, but if you press a biologist about an idea like this, you might encounter a fair bit of uncertainty about the right way to express what has been learned. The biologist might say something like this: "That theory has now been established, especially by genetic evidence. It has been *shown* that mitochondria are descended from free-living bacteria." But you might hear something different, either instead or as well. The biologist might reflect for a moment and say: "Well, all science is entirely provisional. Nothing of any importance is ever conclusively *shown*. This theory about mitochondria is the best one we have right now, but one day it might be overturned." After all, can we ever really be sure about something that was supposed to have happened over one and a half billion years ago?

Scientists often find it difficult to say what it means when a theory has made a transition from being a mere speculation to something routinely taught and assumed in other work. If someone says, "This has been *shown*," that often seems too strong—too unqualified. But if they say, "This is just what we're working with for now," that often seems to understate things. The situation often seems to be something between those two. There are lots of ways of saying something in between; we might say the theory is well supported, or that it has been confirmed by evidence. But when philosophers and scientists have tried to say what support is, and how we might know when a theory has it, the results have often been frustrating. Problems of this kind have long been central to philosophy of science and will be a central thread running through this book.

Something else you will encounter in this book is a lot of disagreement about what kind of theory we should be looking for. A possibility that might come immediately to mind is that we should look for a theory about scientific thinking. Many philosophers have rejected this idea, though, saying that we should seek a logical theory of science. You might not be sure what sort of thing a "logical" theory is. A lot of professional logicians are not sure either. But roughly, the idea is that we might think of a scientific theory as a set of interrelated sentences that make claims about some part of the world (or perhaps just about our experiences). The philosopher might aim to give a description of the relationships that exist between different parts of the theory, and the relationships between the theory and various kinds of evidence we might find–evidence that might support the theory or clash with it. The philosopher might also try to give a description of the logical relationships that can be found between one theory and another.

Philosophers taking this approach tend to be enthusiastic about the tools of mathematical logic. They prize the rigor of their work. This kind of philosophy has often also prompted frustration in people working on the history of science or on how scientific institutions actually operate: the crusty old philosophers seem to be deliberately removing their work from any contact with science as it is actually conducted, perhaps in order to hang on to a set of myths about the perfect rationality of the scientific enterprise. Or perhaps the philosophers want to be sure that nothing too messy will interfere with the endless intellectual games that can be played with imaginary theories expressed in artificial languages.

This kind of logic-based philosophy of science will be discussed in the early chapters of this book. The logical investigations were often very interesting, but ultimately my sympathies lie with those who insist that philosophy of science should have more contact with actual scientific work.

Another view of what we might do in this area is come up with a theory of scientific methods and procedures—perhaps a theory of the "scientific method." The idea of describing a special method that scientists do or should follow is an old one. In the seventeenth century, Francis Bacon and René Descartes, among others, tried to give detailed specifications of how scientists should proceed. Although describing a special scientific method looks like a natural thing to try to do, many people have become skeptical about the idea, especially about the idea of giving anything like a recipe for science. Science, it is often claimed, is too creative and unpredictable a process for there to be a recipe that describes it—this is especially true in the case of great scientists such as

Newton, Darwin, and Einstein. For a long time it was common for science textbooks to have an early section describing "the scientific method," but many textbooks have become more cautious about this.

A distinction that is important all throughout this area is the distinction between *descriptive* and *normative* theories. A descriptive theory is an attempt to describe what actually goes on in some area, or what something is actually like, without making value judgments. A normative theory does make value judgments; it talks about what should go on, or what might be good or bad. Some theories about science are supposed to be descriptive only. But most of the views we will look at do have a normative element, either officially or unofficially. When assessing general claims about science, it is a good principle to constantly ask: "Is this claim intended to be descriptive or normative, or both?"

For some people, the crucial question we need to answer about science is whether or not it is "objective." This term is a slippery one, used to mean a number of very different things. Sometimes objectivity is taken to mean the absence of bias; objectivity is impartiality or fairness. But the term is also often used to express claims about whether the existence of something is independent of our minds. A person might wonder whether there really is an "objective reality"—a reality that exists regardless of how people conceptualize or describe it. We might ask whether scientific theories can ever describe a reality that exists in this sense. Questions like that go far beyond any issue about the absence of bias, and take us into deep philosophical waters.

Several times now I have mentioned fields that are neighbors of the philosophy of science—history of science, sociology of science, and parts of psychology, for example. What is the relation between philosophical theories of science and ideas in these neighboring fields? This question was part of the twentieth-century roller-coaster ride that I referred to earlier. Some people in these neighboring fields thought they had reason to believe that the whole idea of a philosophical theory of science is misguided. They expected that philosophy of science would be replaced by other fields, like sociology. This replacement never occurred. What did happen was that people in these neighboring fields constantly found themselves doing philosophy themselves, whether they admitted it or not. They kept running into questions about truth, about justification, and about the connections between theories and reality. The philosophical problems refused to go away.

Philosophers themselves differ a great deal about what kind of input from these neighboring fields is relevant to philosophy. I think that philosophy of science benefits from a lot of input from other fields. This, too, will be a theme of growing importance as the book goes along.

1.4 *Three Answers, or Pieces of an Answer*

In this section I will introduce three different initial answers to our general questions about how science works. The three ideas can be seen as rivals, or as alternative paths into the problem. But they might instead be considered as pieces of a single, more complicated answer. The problem then becomes how to fit them together.

The first of the three ideas is *empiricism*. Empiricism encompasses a diverse family of philosophical views, and debates within the empiricist camp can be intense. But empiricism is often summarized using something like the following slogan:

Empiricism: The only source of genuine knowledge about the world is experience.

Empiricism, in this sense, is a view about where all knowledge comes from, not just scientific knowledge. How does this help us with the philosophy of science? In general, the empiricist tradition has tended to see the differences between science and everyday thinking as just differences of detail and degree. The empiricist tradition has generally, though not always, tended to construe science in a broad way, and it has tended to approach questions in the philosophy of science from the standpoint of a general theory of thought and knowledge. The empiricist tradition in philosophy has also been largely pro-science; science is seen as the best manifestation of our capacity to investigate and know the world.

Here is a way to use the empiricist principle outlined above to say something about science:

> *Empiricism and Science*: Scientific thinking and investigation have the same pattern as everyday thinking and investigation. In each case, the only source of knowledge about the world is experience. But science is especially successful because it is organized, systematic, and particularly responsive to experience.

The scientific method, insofar as there is such a thing, will then be routinely found in everyday contexts as well. There was no fundamentally new approach to investigation discovered during the Scientific Revolution, according to this view. Instead, Europe was freed from darkness and dogmatism by a few brilliant souls who enabled intellectual culture to "come to its senses."

Some readers are probably thinking that these empiricist principles are empty platitudes. Of course experience is the source of knowledge about the world—what else could be?

For those who suspect that basic empiricist principles are completely trivial, an interesting place to look is the history of medicine. The history of medicine has many examples of episodes where huge breakthroughs were made by people willing to make very basic empirical tests—in the face of much skepticism, condescension, and opposition from people who "knew better." Empiricist philosophers have often used these anecdotes to fire up their readers. Carl Hempel, one of the most important empiricist philosophers of the twentieth century, uses the example of Ignaz Semmelweiss (see Hempel 1966). Semmelweiss worked in a hospital in Vienna in the mid-nineteenth century. He was able to show by simple empirical tests that if doctors washed their hands before delivering babies, the risk of infection in the mothers was hugely reduced. For this radical claim he was opposed and eventually driven from the hospital.

Another example, which can provide a change from the usual case of

Semmelweiss, has to do with the discovery of the role of drinking water in the transmission of cholera.

Cholera was a huge problem in cities in the eighteenth and nineteenth centuries, causing death from terrible diarrhea. Cholera is still a problem whenever there are people crowded together without good sanitation, as it is usually transmitted by the contamination of drinking water. In the eighteenth and nineteenth centuries, there were various theories of how cholera was caused. This was before the discovery of the role of bacteria and other microorganisms in infectious disease. Some thought cholera was caused by foul gases, called *miasmas*, exuded from the ground and swamps. In London, John Snow hypothesized that cholera was spread by drinking water. He mapped the outbreak of one epidemic in London in 1854 and found that it seemed to be centered on a particular public water pump in Broad Street. With difficulty, he persuaded the local authorities to remove the pump's handle. The outbreak immediately ended.

This was an important event in the history of medicine. It was central to the rise of the modern emphasis on clean drinking water and sanitation, a movement that has had an immense effect on human health and well-being. This is also the kind of case that shows the attractiveness of empiricist views.

You might be thinking that we can end the book here. Looking to experience is a guarantee of getting things right. Those who are tempted to think that no problems remain might consider a cautionary tale that follows up the Snow story. This is the tale of brave Dr. Pettenkofer.

Some decades after Snow, the theory that diseases like cholera are caused by microorganisms—the germ theory of disease—was developed in detail by Robert Koch and Louis Pasteur. Koch isolated the bacterium responsible for cholera quite early on. Pettenkofer, however, was unconvinced. To prove Koch wrong, he drank a glass of water mixed with the alleged cholera germs. Pettenkofer suffered few ill effects, and he wrote to Koch saying he had disproved Koch's theory.

Pettenkofer was lucky; Koch was right about what causes cholera. It's not known why Pettenkofer came through unscathed, but the case reminds us that direct empirical tests are no guarantee of success.

Some readers, I said, might be thinking that empiricism is true but too obvious to be interesting. Another line of criticism holds that empiricism is false, because it is committed to an overly simple picture of thought, belief, and justification. The empiricist slogan I gave earlier suggests that experiences pour into the mind and somehow transform themselves into knowledge. But surely the process is more creative than that? In reply, empiricists say they agree that reasoning, including elaborate and creative reasoning, is needed to make sense of what we observe. Still, they insist, the role of experience is fundamental in understanding how we learn about the world rather than fruitlessly spinning our wheels. Many critics of empiricism hold that this is still a mistake; they see it as a hangover from an outdated picture of how belief and reasoning work. That debate will be a recurring theme in this book.

I now turn to the second of the three families of views about the nature of science. This view can be introduced with a quote from Galileo Galilei, one of the heroes of the Scientific Revolution:

> Philosophy is written in this grand book the universe, which stands continually open to our gaze. But the book cannot be understood unless one first learns to comprehend the language and to read the alphabet in which it is composed. *It is written in the language of mathematics*, and its characters are triangles, circles, and other geometric figures without which it is humanly impossible to understand a single word of it; without these, one wanders about in a dark labyrinth. (Galileo [1623] 1990, 237–38; emphasis added)

Putting the point in plainer language, here is the second of the three ideas.

> *Mathematics and Science*: What makes science different from other kinds of investigation, and especially successful, is its attempt to understand the natural world using mathematical concepts and tools.

Is this idea an alternative to the empiricist approach, or something that can be combined with it? An emphasis on mathematical methods has often been used to argue against empiricism. Sometimes this has been because people have thought that mathematics shows us that there must be another route to knowledge beside experience. Mathematics does not depend on experiment and observation, it seems. So although experience is *a* source of knowledge, it can't be the only important source. Alternatively, you might claim that although empiricism is true in broad terms, and all knowledge comes from experience, this tells us nothing about what differentiates science from other areas of human thought. What makes science special is its attempt to quantify phenomena and detect mathematical patterns in the flow of events.

Nonetheless, it is surely sensible to see an emphasis on mathematics as something that can be combined with empiricist ideas. It might seem that Galileo would disagree; Galileo not only exalted mathematics but praised his predecessor Nicolaus Copernicus for making "reason conquer sense [experience]" in his belief that the Earth goes around the sun. But this is a false opposition. In suggesting that the Earth goes around the sun, Copernicus was not ignoring experience, but dealing with apparent conflicts between different aspects of experience. There is no question that Galileo was a very empirically minded person. Observations made using the telescope, for example, were central to his work. So, avoiding the false oppositions, we might argue that mathematics used as a tool within an empiricist outlook is what makes science special.

In this book the role of mathematics will be a significant theme but not a central one—not as central as empiricism. Mathematical tools are not quite as essential to science as Galileo thought. Although mathematics is clearly of huge importance in the development of physics, one of the greatest achievements in all of science—Darwin's *On the Origin of Species* ([1859] 1964)—makes no real use of mathematics. Darwin was not confined to the "dark labyrinth" that Galileo predicted as the fate of nonmathematical investigators. In fact, most (though not all) of the huge leaps in biology that occurred in the nineteenth century occurred without much of a role for mathematics. Biology now contains many mathematical parts, including modern formulations of Darwin's theory of evolution, but this is a more recent development.

Not all science makes much use of mathematics to understand the world. But when it does, mathematics is often so useful that it raises other philosophical problems. Why should the world follow principles that a mathematician's imagination conjures up? It often seems to. The physicist Eugene Wigner published a famous article in 1960 called "The Unreasonable Effectiveness of Mathematics in the Natural Sciences." "Unreasonable" here means mysterious, hard to explain.

The third of the three families of ideas I'll introduce here is newer. Maybe the unique features of science are only visible when we look at scientific communities.

> *Social Structure and Science*: What makes science different from other kinds of investigation, and especially successful, is its unique social structure.

Some of the most important recent work in philosophy of science has explored this idea, but it took the input of historians and sociologists of science to bring philosophical attention to bear on it.

In the hands of historians and sociologists, an emphasis on social structure has often been developed in a way that is strongly critical of the empiricist tradition. Steven Shapin has argued that mainstream empiricism often operates within the fantasy that each person can test hypotheses individually (Shapin 1994). Empiricism is supposed to urge that people be distrustful of authority and go out to look directly at the world. But this, of course, is a fantasy. It is a fantasy in the case of everyday knowledge, and it is an even greater fantasy in the case of science. Almost every move that a scientist makes depends on elaborate networks of cooperation and trust. If each individual insisted on testing everything for themselves, science would never advance beyond the most rudimentary ideas. Cooperation and lineages of transmitted results are essential to science. The case of John Snow—striding like a "lone ranger" up to the Broad Street pump to test his theory about cholera—is very unusual. And even Snow must have been dependent on the testimony of others in his assessment of the state of the cholera epidemic before and after his intervention at the pump.

Trust and cooperation are essential to science. But who can be trusted? Who is a reliable source of data? Shapin argues that when we look closely, we find that a great deal of what went on in the Scientific Revolution had to do with working out new ways of controlling and co-ordinating the actions of groups of people in the activity of research.

Experience is everywhere. The hard thing is working out which kinds of experience are relevant to the testing of hypotheses, and who can be trusted as a source of reliable and relevant reports.

Shapin argues that a good theory of the social organization of science will be a better *theory of science* than empiricist fantasies. But it is also possible to develop theories of how science works that emphasize social organization but are also intended to fit in with a form of empiricism, and I will discuss these later in this book. These accounts of science stress the special balance of cooperation and competition found in scientific communities. People sometimes imagine that seeking credit, and competition for status and recognition, are recent developments in science. But these issues have been important since the Scientific Revolution. The great scientific societies, like the Royal Society of London, came into being quite early—1660 in the case of the Royal Society. One of their tasks was handling the allocation of credit in an efficient way, making sure the right people were rewarded without hindering the free spread of ideas. These societies also functioned to create a community of people who could trust each other as reliable co-workers and sources of data. An empiricist might argue that this social organization made scientific communities uniquely responsive to experience.

In this section I have sketched three families of ideas about how science works and what makes it distinctive. Each idea has sometimes been seen as the starting point for an understanding of science, exclusive of the other two. But it is more likely that they should be seen as pieces of a more complete answer. The first and third themes—empiricism and social structure—are ones we will return to over and over in this book.

1.5 *A Sketch of the Scientific Revolution and What Came Afterward*

Before diving into the philosophical theories, we will take a brief break. Several times already I have mentioned the Scientific Revolution. People, events, and theories from this period carry special weight in discussions of the nature of science. So in this section I will give a historical sketch of the main landmarks in this period. I'll also give a much briefer sketch of how this work fed into the development of science in the centuries that followed.

A good deal of controversy surrounds this period of history, and some historians think that the whole idea of "the Scientific Revolution" is a mistake, as this phrase makes it sound as if there are sharp boundaries between one entirely unique period and the rest of history. But I will use the phrase in the traditional way.

The Scientific Revolution occurred roughly between 1550 and 1700 in Europe. These events are positioned at the end of a series of dramatic social changes, and the Scientific Revolution itself fed into further processes of change. In religion, the Catholic Church had been challenged by the rise of Protestantism. The Renaissance of the fifteenth and sixteenth centuries had included a partial opening of intellectual culture. Populations were growing (recovering from the Black Death), and there was increased activity in commerce and trade. Traditional hierarchies, including intellectual hierarchies, were beginning to show strain. This was a time in which many new, unorthodox ideas were floating around.

The worldview that much of Europe had inherited from the Middle Ages was a combination of Christianity with the ideas of the ancient Greek philosopher Aristotle. The combination is often called the Scholastic worldview, after the "schools" (universities) that developed and defended it. The Earth was seen as a sphere positioned at the center of the universe, with the moon, sun, planets, and stars revolving around it. A detailed model of the motions of these celestial bodies had been developed by the astronomer Ptolemy around 150 CE (the sun was placed between Venus and Mars).

Aristotle's physical theory distinguished "natural" from "violent" or unnatural motion. The theory of natural motions was part of a more general theory of change in which biological development (from acorn to oak, for example) was a central guiding case, and many events were explained using the idea of *purpose*.

Everything on Earth was considered to be made up of mixtures of four basic elements—earth, air, fire, and water—each of which had its own natural tendencies. Objects containing a lot of earth, for example, naturally fall toward the center of the universe, while fire makes things rise. Unnatural motions, such as the motions of projectiles, have an entirely different kind of explanation. Objects in the heavens are made of a fifth element, which is "incorruptible," or unchanging. The natural motion for objects made of this fifth element is circular.

Some versions of this picture included a mechanism (using the term loosely) for the motions of sun, planets, and stars. For example, each body orbiting the Earth might be positioned on a crystalline sphere that revolved around the Earth. Ptolemy's own model was harder to interpret in these terms; Ptolemy is sometimes thought to be most interested in giving a tool for prediction, though interpreters differ on this.

In 1543 the Polish astronomer Nicolaus Copernicus (1473–1543) published a work, *De Revolutionibus*, outlining an alternative picture of the universe (Copernicus [1543] 1992). Others in ancient times had speculated that the Earth might move around the sun instead of vice versa, but Copernicus was the first to give a detailed theory of this kind. In his theory the Earth has two motions, revolving on its axis once a day and orbiting the sun once a year. Copernicus's theory had the same basic placement of the sun, moon, Earth, and the known planets that modern astronomy has. But the theory was made more complicated by his insistence, following Aristotle and Ptolemy, that heavenly motions must be circular. Both the Ptolemaic system and Copernicus's saw most orbits as complex compounds of circles, not single circles. Ptolemy's and Copernicus's systems were about equally complicated, in fact. Writers seem to differ on whether Copernicus's theory was much more accurate as a predictive tool. But there were some famous phenomena that Copernicus's theory explained far better than Ptolemy's. One was the "retrograde motion" of the planets, an apparently erratic motion in which planets seem to stop and backtrack in their motions through the stars.

Copernicus's work aroused interest, but there seemed to be compelling arguments against taking it to be a literally true description of the universe. Some problems were astronomical, and others had to do with obvious facts about motion. Why does an object dropped from a tower fall at the foot of the tower, if the Earth has moved a considerable distance while the object is in flight? Copernicus had entrusted the publication of his 1543 book to a clergyman, Andreas Osiander, who wrote a preface urging that the theory be treated merely as a calculating tool. This became a historically important statement of a view about the role of scientific theories known as *instrumentalism*, which holds that we should think of theories only as predictive tools rather than as attempts to describe the hidden structure of nature.

The situation was changed dramatically by Galileo Galilei (1564–1642), working in Italy in the early years of the seventeenth century. Galileo made the case for the literal truth of the Copernican system, as opposed to its mere usefulness. Galileo used telescopes (which he did not invent but did improve) to look at the heavens and found a multitude of phenomena that contradicted Aristotle and the Scholastic view of the world. He also used a combination of mathematics and experiment to begin the formulation of a new science of motion that would make sense of the idea of a moving Earth and explain familiar facts about dropped and thrown objects. Galileo's work eventually aroused the ire of the pope; he was forced by the Inquisition to recant his Copernican beliefs and spent his last years under house arrest. (Galileo was treated lightly in comparison with Giordano Bruno, whose refusal to disown his unorthodox speculations about the place of the Earth in the universe led to his being burned at the stake in Rome, for heresy, in 1600.)

Galileo remained wedded to circular motion as astronomically fundamental. The move away from circular motion was made by Johannes Kepler (1571–1630), a mystical thinker who combined Copernicanism with an obsession with finding mathematical harmony (including musical chords) in the structure of the heavens. Kepler's model of the universe, also developed around the start of the seventeenth century, had the Earth and other planets moving in *ellipses*, rather than circles, around the sun. This led to much simplification and better predictive accuracy.

So far I have mentioned only changes in astronomy and related areas of physics, and I have taken the discussion only to the early part of the seventeenth century. Part of what makes this initial period so dramatic is the removal of the Earth from the center of the universe, an event laden with symbolism. Another field that changed in the same period is anatomy. In Padua, Andreas Vesalius (publishing, like Copernicus, in 1543) began to free anatomy from its dependence on ancient authority (especially Galen's views) and set it on a more empirical path. Influenced by Vesalius's school, William Harvey achieved the most famous break-through in this period, establishing in 1628 the circulation of blood and the role of the heart as a pump.

The mid-seventeenth century saw the rise of a general and ambi-tious new theory about matter: *mechanism*. The mechanical view of the world combined ideas about the composition of things with ideas about causation and explanation. According to mechanism, the world is made up of tiny "corpuscles" of matter, which interact only by local physical contact. Ultimately, good explanations of physical phenomena should only be given in terms of mechanical interactions. The universe was to be understood as operating like a mechanical clock.

Some, including René Descartes (1596–1650), thought that immaterial souls and a Christian God must be posited as well as physical corpuscles. Though many figures in the Scientific Revolution held religious views that were at least somewhat unorthodox, they were generally not looking for a showdown with mainstream religion. Most of the "mechanical phi-losophers" retained a role for God in their overall pictures of the world. (If the world is a clock, who set it in motion?) However, the idea of dropping souls, God, or both from the picture was sometimes considered.

In England, Robert Boyle (1627–1691) and others embedded a version of mechanism into an organized and well-publicized program of research that urged systematic experiment and the avoidance of unempirical speculation. In the mid-seventeenth century we also see the rise of sci-entific societies in London, Paris, and Florence. These were intended to organize the new research and break the institutional monopoly of the (often conservative) universities.

This period ends with the work of Isaac Newton (1642–1727). In 1687 Newton published his *Principia*, which gave a unified mathematical treatment of motion both on Earth and in the heavens. Newton showed why Kepler's elliptical orbits were the inevitable outcome of the force of gravity operating between heavenly bodies, and he vastly improved the ideas about motion on Earth that Galileo and others had pioneered. So impressive was this work that for hundreds of years Newton was seen as having essentially completed those parts of physics. Newton also did immensely influential work in mathematics and optics, and he suggested the way forward in fields like chemistry. In some ways Newton's physics was the culmination of the mechanical worldview, but in other ways it was "post-mechanical," since it posited some forces (gravity, most importantly) that were hard to interpret in mechanical terms.

By the end of the seventeenth century, the Scholastic worldview had been replaced by a combination of Copernicanism and a form of mechanism. As far as method is concerned, a combination of experiment and mathematical analysis had triumphed (though people disagreed about the nature of the triumphant combination). This ends the period usually referred to as the Scientific Revolution. The events described above fed into further changes, both intellectual and political. Chemistry began a period of rapid development in the mid to late eighteenth century, a period sometimes called the Chemical Revolution. The work of Antoine Lavoisier, especially his description of oxygen and its role in combustion, is often taken to initiate this revolution, though it was in the nineteenth century, with the work of John Dalton, Dmitri Mendeleev, and others, that the basic features of modern chemistry, like the periodic table of elements, were established.

In the eighteenth century the philosophers of the French Enlightenment hoped to use science and reason to sweep away ignorance and superstition, along with oppressive religious and political institutions. The intellectual movements leading to the American and French Revolutions in the late eighteenth century were much influenced by currents of thought in science and philosophy. These included empiricism, mechanism, the inspiration of Newton, and a general desire to understand mankind and society in a way modeled on the understanding of the physical world achieved during the Scientific Revolution.

Moving back to how things went within science, Carl Linnaeus systematized biological classification in the eighteenth century, but it was the nineteenth century that saw dramatic developments in biology. These developments include the theory that organisms are comprised of cells, Darwin's theory of evolution, the germ theory of disease, and the work by Gregor Mendel on inheritance across generations that laid the foundation for genetics. In the twentieth century, the "vitalist" idea that there might be a fundamental difference between living and non-living matter subsided as biological processes began to be understood first in chemical terms and then at the molecular level.

Many of the most important development in physics during the nineteenth century concerned electricity and magnetism, and the unification of these phenomena in *electromagnetism*. Michael Faraday and James Clerk Maxwell, the two most important figures in that field, introduced a way of thinking about the world in terms of dispersed, extended fields rather than (or as well as) particles. Massive technological change followed as electricity and electromagnetic fields were harnessed in motors, lights, and radio transmission. Around the turn of the twentieth century, physics was transformed entirely with quantum mechanics and Einstein's theories of relativity. The revolution in physics at this time, which was immensely disruptive and surprising, had effects also on philosophy. Many of the philosophical ideas encountered in the early chapters of this book were profoundly influenced by the fall of Newton's physics despite centuries of success, and the rise of predictively powerful but conceptually puzzling theories of space, time, and matter.

Further Reading and Notes

Starting with some general reference works, Blackburn's *Oxford Dictionary of Philosophy* (2008) is a very useful book and amusing to browse through. *The Stanford Encyclopedia of Philosophy* is a high-quality free online resource, much better than most Internet sources for information about philosophers and philosophical ideas. Most of the topics in this book have a good *Stanford Encyclopedia* article. I cite a few of these, but there are many more.

For the debates about anthropology as science, see website [1].

Archibald (2014) covers the history of ideas about symbiosis in cell biology, including the role of Lynn Margulis. (Her classic article of 1967 was published under her married name at that time, Lynn Sagan.)

Evans (1973) revisits the case of Dr. Pettenkofer and cholera, which has more twists and turns than is often realized. The Broad Street epidemic, as Snow later admitted, was also becoming less intense even before he removed the handle of the water pump. For a history of medicine covering the entire world, see Porter's *The Greatest Benefit to Mankind* (1998).

There are many good books on the Scientific Revolution, each with a different emphasis. Cohen (1985) is a classic and very helpful on the physics. Henry (1997) is concise and thorough, with an excellent chapter on mechanism. Toulmin and Goodfield's *The Fabric of the Heavens* (1962), quite an old book now, is my favorite. It focuses on the conceptual foundations underlying the development of scientific ideas. (This is the first of three books by Toulmin and Goodfield on the history of science; the second, *The Architecture of Matter* [1982], is also relevant here.) Galileo's comment about making "reason conquer sense" is found in his *Dialogue Concerning the Two Chief World Systems* (1632), translated by Stillman Drake.

Kuhn's *The Copernican Revolution* (1957) is rather difficult but a classic, focused on the early stages, as the title suggests. Shapin's *The Scientific Revolution* (1996) is not a good introduction to the Scientific Revolution but is an interesting book anyway. He is the person I had in mind when I said that some people think the Scientific Revolution does not really exist. His book begins: "There was no such thing as the Scientific Revolution, and this is a book about it." There are several good books that focus on particular personalities. Koestler (1968) is fascinating on Kepler, and Sobel (1999) is good on Galileo (and his daughter, a nun who led a tough life). The standard biography of the extremely strange Isaac Newton, by Robert Westfall, comes in both long (1980) and short (1993) versions.

Faraday, Maxwell, and the Electromagnetic Field, by Nancy Forbes and Basil Mahon (2014), is readable and engaging (though rather full of adulation). For Einstein's theories, which have particular importance to the development of the philosophy of science, see his own popular summary, *Relativity: The Special and the General Theories* ([1916] 2015). It is not easy but worth the effort. An excellent book about another side of twentieth-century physics and its consequences is Richard Rhodes's *The Making of the Atomic Bomb* (1987).

Chapter 2

Empiricism

2.1 *The Empiricist Tradition*

Empiricism is a family of ideas that have probably shaped philosophical thinking about science more than any other. In addition, a particular version of empiricism that arose early in the twentieth century set in motion the sequence of debates I will describe chronologically in the first half of this book.

In the opening chapter I said that empiricism is often summarized with the claim that the only source of knowledge is experience. This idea has a long history, but the most important stage in the development of empiricist philosophy was in the seventeenth and eighteenth centuries, with the work of John Locke, George Berkeley, and David Hume. These "classical" forms of empiricism were based upon theories about the mind and how it works. Their view of the mind is sometimes called "sensationalist." Sensations, like patches of color and sounds, appear in the mind and are all the mind has access to. The role of thought is to track and respond to patterns in these sensations. Using a phrase that was unknown then, but useful anyway, we could say that classical empiricism saw the mind largely as a pattern-recognition device.

Both during these classical discussions and more recently, a problem for empiricism has been a tendency to lapse into *skepticism*, the idea that we cannot know anything, or can only know much less than is usually supposed, about the world and its workings. There are many kinds of skepticism, but two are especially important here. One is *external world skepticism*, which questions whether we can ever know anything about a physical world that might lie behind the flow of sensations we receive. The second form, made vivid by Hume, is *inductive skepticism*: why do we have reason to think that the patterns found in past experience will also hold in the future?

Empiricists have often shown a surprising willingness to throw in the towel when faced with external world skepticism. (Hume threw in the towel on both kinds, but that is unusual.) Quite a few empiricists have said that they don't care about the possibility that there might be real things lying behind the flow of sensations. It's only the sensations that we have any dealings with. Maybe it makes no sense even to try to think about objects lying behind sensations—perhaps our concept of "the world" is just a concept of a patterned collection of sensations. This view is sometimes called "phenomenalism." During the nineteenth century, phenomenalist views were quite popular, and their oddity was treated with nonchalance. John Stuart Mill, a leading English philosopher and political theorist, once said that matter may be defined as "a Permanent Possibility of Sensation" (1865, 183). Ernst Mach, an Austrian physicist and philosopher who influenced many people, including Einstein, illustrated his phenomenalist view by drawing a picture of the world as it appeared through his left eye (see fig. 2.1; the shape in the lower right part of the image is his elegant mustache). All that exists is a collection of observer-relative sensory phenomena like these.

I hope phenomenalism looks strange to you, despite its eminent proponents. It is a strange idea. But empiricists have often found themselves backing into views like this, and when they arrive they sometimes say they feel at home there. This is partly because empiricists have often tended to think of the mind as confined behind a "veil of ideas" or sensations. The mind has no access to anything outside the veil. Many philosophers, including me, agree that this picture of the mind is a mistake. But it is not easy to set up an empiricist view that entirely avoids the influence of this picture.

In discussions of the history of philosophy, it is common to talk of a battle in the seventeenth and eighteenth centuries between the "rationalists" and the "empiricists." Rationalists such as Descartes and G. W. Leibniz believed that pure reasoning can be a route to knowledge that does not depend on experience. Mathematics seemed to be a compelling example of this kind of knowledge. Empiricists like Locke and Hume insisted that experience is our only way of finding out what the world is like. In the late eighteenth century, a sophisticated intermediate position was developed by the German philosopher Immanuel Kant. Kant argued that all our thinking involves a subtle interaction between sensory experience and preexisting mental structures that we use to make sense of experience. Concepts such as space, time, and causation cannot be derived from experience, because a person must already have

Figure 2.1. "The assertion, then, is correct that the world consists only of our sensations." (Mach 1897, 10)

these concepts in order to use experience to learn about the world. Kant also held that mathematics gives us genuine knowledge but does not require experience for its justification.

As I said earlier, in the history of philosophy the term "rationalism" is often used for a view that opposes empiricism. In the more recent discussions of science that we are concerned with here, however, the term is generally not used in that way. (This can be a source of confusion; see the glossary.) The views referred to as rationalist in the twentieth century were often also forms of empiricism; the term "rationalism" was often used in a broad way, to indicate confidence in the power of human reason.

Despite various problems, empiricism has been an attractive set of ideas for many philosophers. Empiricism has often also had a particular kind of impact on discussions outside of philosophy. Making a sweeping generalization, it is fair to say that the empiricist tradition has tended to be (1) pro-science, (2) worldly rather than religious, and (3) politically moderate or liberal (though these political labels can be hard to apply across centuries). Hume, Mill, and Bertrand Russell are examples of all three parts of this tendency. Of the three elements of my generalization, religion is the one that has the most exceptions—Berkeley was a bishop, for example. But on the whole it is fair to say that empiricist ideas have tended to be the allies of a practical, scientific, down-to-earth outlook on life.

2.2 *The Vienna Circle*

A new form of empiricism developed in Europe after World War I. The movement was established by a group of people who were scientifically oriented and who disliked much of what was happening in philosophy. They also thought they could avoid many of the problems with traditional forms of empiricism. This group has become known as the "Vienna Circle" (though their own name for the group paid homage to Mach, who did the drawing that appears as fig. 2.1). The Vienna Circle was established by Moritz Schlick and Otto Neurath. It was based, as you might expect, in Vienna, Austria. Another member of the group who was central to the development of its ideas was Rudolf Carnap.

The usual name for the view the Vienna Circle developed is "logical positivism." (The term "positivism" derives from the nineteenth-century scientific philosophy of Auguste Comte.) The view is sometimes called "logical empiricism" instead, though other people use this pair of terms to mark a distinction within the movement, saying "logical positivism" for an earlier, more extreme form of the view and "logical empiricism" for a later, more moderate version. I will follow that usage.

It is worth spending some time describing the unusual intellectual and historical context in which logical positivism developed. In particular, it is worth paying attention to what the logical positivists were against. The logical positivists were inspired by developments in science in the early years of the twentieth century, especially the work of Einstein. They also thought that developments in logic, mathematics, and the philosophy of language had shown a way to put together a new kind of philosophy. Some traditional problems would be solved by this approach, while others would be rejected as meaningless. Logical positivist views about language were influenced by the early ideas of Ludwig Wittgenstein ([1922] 1988). Wittgenstein was an enigmatic, charismatic, and eccentric philosopher of logic and language who was not an empiricist at all. Some would say that the positivists adapted Wittgenstein's ideas, others that they misinterpreted him.

Though they admired some philosophers, the logical positivists were distressed with much of what had been going on in philosophy. In the years after Immanuel Kant's death in 1804, philosophy had seen the rise of a number of systems of thought that the logical positivists found pretentious, obscure, dogmatic, and politically harmful. A central villain was G. W. F. Hegel, who had a huge influence on nineteenth-century thought. Hegel was famous for his work on the relation between philosophy and history. He thought that human history as a whole was a process in which a "world spirit" gradually reached consciousness of itself. For Hegel, individuals are less important than the state as a whole, especially the role of the state in the grand march of historical progress. These ideas were often taken to support nationalism. Hegel's was an "idealist" philosophy, since it held that reality is in some sense spiritual or mental. But this is not a view in which each person's reality is made up in some way by that person's ideas. Rather, a single reality *as a whole* is said to have a spiritual or rational character. This view is sometimes called "absolute idealism."

Hegel's influence bloomed and then receded in continental Europe. As it receded in continental Europe, in the later nineteenth century, it bloomed in England and America. Absolute idealism is a good example of what logical positivism was against. Sometimes the positivists would disparagingly dissect especially obscure passages from this literature. Hans Reichenbach (who was not part of the original Vienna Circle but was a close ally) began his book *The Rise of Scientific Philosophy* (1951) with a quote from Hegel's most famous work on philosophy and history: "Reason is substance, as well as infinite power, its own infinite material underlying all the natural and spiritual life; as also the infinite form, that which sets the material in motion." Reichenbach lamented that if you are a philosophy student, you might initially think it is your fault that you do not understand Hegel's impressive-looking words. You will then work away until finally it seems undeniable to you that reason, indeed, really is substance, as well as infinite power. . . . Reichenbach, in contrast, says it is Hegel's fault that the passage seems to make no sense. Whatever factual meaning the claim might be intended to convey has been smothered with misused language. The absolute idealist emperor has no clothes, and we all need to realize this and not be intimidated by elaborate words.

Another philosopher who came to seem an especially important rival to logical positivism was Martin Heidegger. A moment ago I gave a quick summary of Hegel's ideas. It is much harder to do that for Heidegger, especially for his most influential work, *Being and Time* (1927). Heidegger is sometimes categorized as an existentialist. Perhaps he is the most famously difficult and obscure philosopher who has ever lived (so far). I will borrow the summary reluctantly given by Thomas Sheehan in the entry for Heidegger in the *Routledge Encyclopedia of Philosophy* (1998): "He argues that mortality is our defining moment, that we are thrown into limited worlds of sense shaped by our being-towards-death, and that finite meaning is all the reality we get." Perhaps this does not get us all the way to a straightforward summary. Simplifying even more, Heidegger held that we must understand our lives as based, first and foremost, upon practical coping with the world rather than knowledge of it. All our experience is affected by the awareness that we are traveling toward death. The best thing we can do in this situation is stare it in the face and live an "authentic" life.

The emphasis on practical coping in Heidegger's view, and the resulting contrasts with more traditional philosophies, are quite interesting. He criticizes attempts to explain features of human life by fitting them into a picture governed by our scientific knowledge. But Heidegger combined these ideas with tremendously convoluted discussions of Being and—notoriously—the nature of "Nothing." Heidegger also had one point in common with some (though not all) absolute idealists: his opposition to liberal democratic political ideas.

Heidegger was seen as an important rival by the logical positivists. Carnap gave humorous logical dissections of Heidegger's discussions of Nothing in his lectures. Logical positivism was a plea for Enlightenment values, in opposition to mysticism, Romanticism, and nationalism. The positivists championed reason over the obscure, the logical over the intuitive. The logical positivists were also internationalists, and liked the idea of a universal and precise language that everyone could use to communicate clearly. Otto Neurath was the member of the group with the strongest political and social interests. He and various others in the group could be described as democratic socialists. They also had a keen interest in some movements in art and architecture at the time, such as the Bauhaus movement. They saw this work as assisting the development of a scientific, internationalist, and progressive outlook on society.

The Vienna Circle flourished from the mid-1920s to the mid-1930s. Logical positivist ideas were imported into England by A. J. Ayer in *Language, Truth, and Logic* (1936), a vivid and readable book that conveys the excitement of the time. Under the influence of logical positivism and the philosophy of G. E. Moore and Bertrand Russell, English philosophy abandoned absolute idealism and returned to a more empiricist emphasis.

In continental Europe the story turned out differently. For we have now, remember, reached the 1930s. The development of logical positivism ran straight into the rise of Adolf Hitler.

Many of the Vienna Circle had socialist leanings, some were Jewish, and there were certainly no Nazis. So the logical positivists were persecuted by the Nazis, to varying degrees. The Nazis encouraged and made use of pro-German, anti-liberal philosophers, who also tended to be obscure and mystical. Martin Heidegger joined the Nazi party in 1933 and remained a member throughout the war. Many logical positivists fled Europe, especially to the United States. Schlick, unfortunately, did not. He was murdered by a psychopathic former student in 1936. (At his trial, the student said that Schlick's philosophy had undermined his moral restraints.) The logical positivists who did make it to the United States were responsible for a great flowering of American philosophy in the years after World War II. These include Rudolf Carnap, Hans Reichenbach, Carl Hempel, and Herbert Feigl. In the United States, the strident voice of logical positivists was moderated. This was partly because of criticisms of their ideas—criticisms from the side of those who shared their general outlook. But the moderation was no doubt due in part to the different intellectual and political climate in the United States. Austria and Germany in the 1930s had been an unusually intense environment for doing philosophy.

2.3 *Central Ideas of Logical Positivism*

Earlier empiricist views were based on views about the mind and perception. Logical positivism, in contrast, was based in large part on theories about language—especially about what language can and can't express. Perhaps their central idea was the *verifiability theory of meaning.*

Here is how the theory was often put: the meaning of a sentence consists in its method of verification. That formulation might sound strange (it always has to me). Here is a formulation that sounds more natural: knowing the meaning of a sentence is knowing how to verify it. And here is a key application of the principle: if a sentence has no possible method of verification, it has no meaning.

By "verification," the positivists meant verification by means of observation. Observation in all these discussions is construed broadly, to include many kinds of sensory experience. And "verifiability" is not the best word for what they meant. A better word would be "testability." This is because testing is an attempt to work out whether something is true or false, and that is what the positivists had in mind. The term "verifiable" generally only applies when you are able to show that something is true. It would have been better to call the theory "the testability theory of meaning." Sometimes the logical positivists did use that phrase, but the more standard name is "verifiability theory," or just "verificationism."

Verificationism is a strong empiricist principle; experience is the source of meaning when we speak and write, as well as the only source of knowledge. Note that verifiability here refers to verifiability in principle, not in practice. There was some dispute about which hard-to-verify claims are really verifiable in principle. It is also important that conclusive verification or testing was not required. There just had to be the possibility of finding observational evidence that would count for or against the proposition in question. In addition, the verifiability theory was only supposed to apply to a particular kind of meaning, the kind seen when a person is trying to state something about the world—rather than issue a command, or express an emotional response, for example. This was sometimes called "factual meaning."

In the early days of logical positivism, the claim was that in principle one could translate all factually meaningful sentences into other sentences that referred only to possible observations and the patterns connecting them. This program of translation was soon abandoned as too extreme. But the verifiability theory was retained after the program of translation had been dropped. The logical positivists used the verifiability principle as a philosophical weapon. Scientific discussion, and most everyday discussion, consists of verifiable and hence meaningful claims. Some other parts of language are not even intended to have factual meaning, so they fail the verifiability test but in a harmless way. Included here are poetic language, expressions of emotion, and so on. But there are also parts of language that are supposed to have factual meaning—are supposed to say something about the world—and fail to do so. For the logical positivists, this includes most traditional philosophy, much of ethics, and theology as well.

At this point you might be wondering about that long-standing problem for empiricist views: mathematical statements. To describe the logical positivist response to this problem I need to introduce a second part of their view of language. This is the distinction between *analytic* and *synthetic* sentences.

Some sentences are true or false simply in virtue of the meaning of the words within them, regardless of how the world happens to be; these are analytic. A synthetic sentence is true or false in virtue of both the meaning of the words in the sentence and how the world actually is. "All bachelors are unmarried" is a standard example of an analytically true sentence. "All bachelors are bald" is an example of a synthetic sentence, in this case a false one. Analytic truths are, in a sense, empty truths, with no factual content. Their truth has a kind of necessity, but only because they are empty. Analytic sentences are not supposed to be covered by the verifiability theory of meaning; they are another example of sentences with a different kind of meaning that is not "factual."

This distinction had been around, in various forms, since at least the eighteenth century. The terminology "analytic/synthetic" was introduced by Kant. Although the distinction itself looks uncontroversial, it can be made to do real philosophical work. Here is some of that work: the logical positivists claimed that all of mathematics and logic is analytic. For logical positivism, mathematical propositions do not describe the world; they merely record our decision to use symbols in a particular way. Synthetic claims about the world can be expressed using mathematical terms, such as when it is claimed that Jupiter has seventy-nine moons. But proofs and investigations within mathematics itself are analytic. This might seem unlikely, because some statements and proofs in mathematics are so surprising and certainly look significant. For example, there are infinitely many prime numbers—that claim does not look empty at all. But the logical positivists insisted that once we break down any mathematical proof into small steps, each step will be trivial and unsurprising.

Earlier philosophers in the rationalist tradition had claimed that some things can be known a priori, this means known *independently of experience*. Logical positivism held that the only things that seem to be knowable a priori are analytic and hence empty of factual content. A remarkable episode in the history of science is important here. For many centuries, the geometry of the ancient Greek mathematician Euclid was regarded as a shining example of real and certain knowledge. (Euclid, incidentally, also proved that there are infinitely many prime numbers.) Immanuel Kant, inspired by the immensely successful application of Euclidean geometry to nature in Newtonian physics, claimed that Euclid's geometry (along with the rest of mathematics) is both synthetic and knowable a priori. In the nineteenth century, mathematicians worked out some alternative geometrical systems to Euclid's, but they did so as a mathematical exercise, not as an attempt to describe how lines, angles, and shapes work in the actual world. Early in the twentieth century, however, Einstein's revolutionary work in physics showed that a non-Euclidean geometry *is* true of our world. The logical positivists were very impressed by this development, and it guided their analysis of mathematical knowledge. The positivists insisted that pure mathematics is analytic, and they broke geometry into two parts. One part is purely mathematical, analytic, and says nothing about the world. It merely describes possible geometrical systems. The other part of geometry is a set of synthetic claims about which geometrical system applies to our world.

Another part of their view of language—a part that brings us closer to issues about science—is a distinction they made between *observational* and *theoretical* language. There was uncertainty about exactly how to set up this distinction. Usually it was seen as a distinction applied to individual terms. "Red" is in the observational part of language, and "electron" is in the theoretical part. There was also a related distinction at the level of sentences. "The rod in front of me is glowing red" is observational, while "Helium atoms each contain two electrons" is theoretical. A more important question was where to draw the line. Schlick thought that only terms referring to sensations were observational; everything else was theoretical. Here Schlick stayed close to traditional empiricism. Neurath thought this was a mistake and argued that many terms referring to ordinary physical objects are in the observational part of language. For Neurath, scientific testing must not be understood in a way that makes it private to the individual. Only observation statements about physical objects can be the basis of public or "intersubjective" testing.

Carnap came to think that there are lots of acceptable ways of marking out a distinction between the observational and theoretical parts of language; one can use whichever is convenient for the purposes at hand. This was the start of a more general move that Carnap made toward a view based on the "tolerance" of alternative linguistic frameworks.

This analysis of language provided the background ideas for the logical positivist philosophy of science. Science was seen as a more complex and sophisticated version of the same sort of thinking, reasoning, and problem-solving that we find in everyday life, and completely unlike the meaningless blather of traditional philosophy.

Though traditional philosophy was seen as largely a waste of time, the logical positivists did think there were some real tasks for philosophers to do. These tasks were mostly concerned with logic. They saw logic as the main tool for philosophy, including philosophical discussion of science. The most useful thing that philosophers can do is give logical analyses of how language, mathematics, and science work.

Here we should distinguish between two kinds of logic (this discussion will be continued in chapter 3). Logic in general is the attempt to give an abstract theory of what makes some arguments compelling and reliable. Deductive logic is the most familiar kind of logic, and it describes patterns of argument that transmit truth with certainty. These are arguments with the feature that if the premises of the argument are true, the conclusion must be true. Impressive developments in deductive logic had been under way since the late nineteenth century and were still going on at the time of the Vienna Circle.

The logical positivists also believed in a second kind of logic, a kind that was (and is) much more controversial. This is *inductive* logic. Inductive logic was supposed to be a theory of arguments that provide support for their conclusions but do not give the kind of guarantee found in deductive logic.

From the logical positivist point of view, developing an inductive logic was of great importance. Hardly any of the reasoning about the world that we encounter in everyday life and science carries the kind of guarantee found in deductive logic. Even the best kind of evidence we can find for a scientific theory is not completely decisive. There is always the possibility of error, but that does not stop some claims in science from being supported by evidence. The logical positivists accepted and embraced the fact that error is always possible. Although some critics have misinterpreted them on this point, the logical positivists did not think that science ever reaches absolute certainty.

The logical positivists saw the task of logically analyzing science as sharply distinct from any attempt to understand science in terms of its history or psychology. Those are empirical disciplines, and they involve a different set of questions from those of philosophy. A terminology standardly used to express the separations between different approaches here was introduced by Reichenbach, who distinguished between the "context of discovery" and the "context of justification." That terminology is not very helpful, because it suggests that the distinction has to do with before versus after. It might seem that the point being made is that discovery comes first and justification comes afterward. That is not the point (though the logical positivists were not completely clear on this). The distinction, instead, is between the study of the logical structure of science and the study of all the historical and psychological aspects of science. Logical positivism tended to dismiss the relevance of fields like history and psychology to the philosophy of science. In time, this came to seem a big mistake.

Let us put all these ideas together and look at the picture of science that results. Logical positivism was a revolutionary, uncompromising version of empiricism, based largely on a theory of language. The aim of science—and the aim of everyday thought and problem-solving as well—is to track and anticipate patterns in experience. As Schlick once put it, "what every scientist seeks, and seeks alone, are . . . the rules which govern the connection of experiences, and by which alone they can be predicted" (1932–33, 44). We can make rational predictions about future experiences by attending to patterns in past experience, but we never get a guarantee. We could always be wrong. There is no alternative route to knowledge besides experience; when philosophy has tried to find such a route, it has lapsed into meaninglessness.

During the early twentieth century, other versions of empiricism were being developed as well. One was *operationalism*, which was introduced by a physicist, Percy Bridgman (1927). Operationalism held that scientists should use language in such a way that all theoretical terms are tied closely to direct observational tests. This is akin to logical positivism, but it was expressed more as a proposed tightening up of scientific language than as an analysis of how all science already works.

In the latter part of the twentieth century, an image of the logical positivists developed in which they were seen as stodgy, conservative, unimaginative science worshipers. Their pro-science stance has been seen as antidemocratic, or aligned with repressive political ideas. This is very unfair, given their actual political interests and activities. Later we will see how ideas about the relation between science and politics changed through the twentieth century in a way that made this interpretation possible. The accusation of stodginess is another matter; the logical positivists' writings were indeed often dry and technical. Still, even the driest of their ideas were part of a remarkable program that aimed at a massive, transdisciplinary, intellectual housecleaning. And their version of empiricism was organized around an ideal of intellectual flexibility as a mark of science and rationality. We see this in a famous metaphor used by Neurath. He said that in our attempts to learn about the world and improve our ideas, we are "like sailors who have to rebuild their ship on the open sea." The sailors replace pieces of their ship plank by plank, in a way that eventually results in major changes, but is constrained by the need to keep the ship afloat during the process.

2.4 *Problems and Changes*

Logical positivist ideas were always in a state of flux and were subject to many challenges. One set of problems was internal to the program. There was considerable difficulty in getting a good formulation of the verifiability principle. It turned out to be hard to formulate the principle in a way that would exclude all the obscure traditional philosophy but include all of science. Some of these problems were almost comically simple. For example, if "Metals expand when heated" is testable, then

"Metals expand when heated and the Absolute Spirit is perfect" is also testable. If we could empirically show the first part of the claim to be false, then the whole claim would be shown false, because of the logic of statements containing "and." (If *A* is false, then *A&B* must be false too, no matter what *B* is.) Patching this hole led to new problems elsewhere; the whole project was quite frustrating (Hempel 1965, chap. 4). The attempt to develop an inductive logic also ran into serious trouble, a topic that will be covered in the next chapter.

Other criticisms were directed not at the details but at more fundamental ideas. I will spend some time on one of these, a criticism presented in an article that had a huge influence on philosophy in the middle of the twentieth century: W. V. Quine's "Two Dogmas of Empiricism" (1951a).

Quine argued for a *holistic* theory of testing, and he used this to motivate a holistic theory of meaning as well. In describing the view, first I should say something about holism in general. Many areas of philosophy contain views that are described using the term "holism." A holist argues that you cannot understand a particular thing without looking at its place in a larger whole. In the case we are concerned with here, holism about testing says we cannot test a single hypothesis or sentence in isolation. Instead, we can only test complex networks of claims and assumptions. This is because only a complex network of claims and assumptions makes definite predictions about what we should observe.

Let us look more closely at the claim that individual claims about the world cannot be tested in isolation. The idea is that in order to test one claim, you need to make assumptions about many other things. Often these will be assumptions about measuring instruments, the circumstances of observation, the reliability of records and other observers, and so on. Whenever you think of yourself as testing a single idea, what you are really testing is a long, complicated *conjunction* of statements ($p \& q \& r \& \ldots$); it is the whole conjunction that gives you a definite prediction. If a test has an unexpected result, then something in that conjunction is false, but the test itself does not tell you where the error is.

For example, suppose you want to test the hypothesis that high air pressure is associated with fair, stable weather. You make a series of observations, and what you seem to find is that high pressure is instead associated with unstable weather. It is natural to suspect that your original hypothesis was wrong, but there are other possibilities as well. It might be that your barometer does not give reliable measurements of air pressure. There might also be something wrong with the observations made (by you or others) of the weather conditions themselves. The unexpected observations are telling you that something is wrong, but the problem might lie with one of your background assumptions, not with the hypothesis you were trying to test.

Some parts of this argument are convincing. It is true that only a network of claims and assumptions, not a single hypothesis alone, tells us what we should expect to observe. The failure of a prediction will always have a range of possible explanations. In that sense, testing is indeed holistic. But this leaves open the possibility that we might often have good reasons to lay the blame for a failed prediction at one place rather than another. In practice, science often seems to have effective ways of working out where to lay the blame. Giving a philosophical theory of these decisions is a difficult task, but the mere fact that failed predictions always have a range of possible explanations does not settle the holism debate.

These holist arguments were very influential, though. Quine, who sprinkled his writings with deft analogies and dry humor, argued that mainstream empiricism had been committed to a badly simplistic view of testing. We must accept, as Quine said in a famous metaphor, that our theories "face the tribunal of sense-experience . . . as a corporate body" (1951a, 38). In simpler language, from another paper: "Science is a unified structure, and in principle it is the structure as a whole, and not its component statements one by one, that experience confirms or shows to be imperfect" (1951b, 72). Logical positivism, he said, must be replaced with a holistic version of empiricism. But there is a puzzle here. The logical positivists already accepted that testing is holistic in the sense described above. Here is Herbert Feigl: "No scientific assumption is testable in complete isolation. Only whole complexes of inter-related hypotheses can be put to the test" (1943, 16). Carnap had been saying the same thing (1937, 318). We can even find statements like this in Ayer's *Language, Truth, and Logic* (1936).

Quine did recognize Pierre Duhem, a French physicist and philosopher, as someone who had argued for holism about testing before him. (Holism about testing is often called the "Duhem-Quine thesis.") But how could it be argued that logical positivists had dogmatically missed this important fact, when they repeatedly expressed it in print? Regardless, many philosophers agreed with Quine that logical positivism had made a bad mistake about testing in science.

Though the history of the issue is strange, it might be fair to say this: although the logical positivists officially accepted a holistic view about testing, they did not appreciate the significance of the point. The verifiability principle seems to suggest that you can test sentences one at a time. It seems to attach a set of observable outcomes of tests to each sentence in isolation.

Strictly, the positivists generally held that these observations are only associated with a specific hypothesis against a background of other assumptions. But then it seems questionable to associate the test results solely with the hypothesis itself and not the other assumptions. Quine made the consequences of holism about testing very clear. He also drew conclusions about language and meaning; given the link between testing and meaning asserted by logical positivism, holism about testing leads to holism about meaning. And holism about meaning causes problems for many logical positivist ideas.

The version of holism that Quine defended in "Two Dogmas" was an extreme one. It included an attack on the one idea in the previous section that you might have thought was completely safe: the analytic/synthetic distinction. Quine argued that this distinction does not exist; this is another unjustified dogma of empiricism.

Here again, some of Quine's arguments were directed at a version of the analytic/synthetic distinction that the logical positivists no longer held. Quine said that the idea of analytic truth was intended to treat some claims as *immune to revision* (as long as ordinary lapses in reasoning are not an issue), and he argued that actually no statement is immune to revision. But Carnap already accepted that analytic statements can be revised, though they are revised in a special way. A person or community can decide to drop one whole linguistic and logical framework and adopt another. Against the background provided by a given linguistic and logical framework, some statements will be analytic and hence not susceptible to empirical test. But we can always change frameworks. By the time that Quine was writing, Carnap's philosophy was based on a distinction between changes made *within* a linguistic and logical framework, and changes *between* these frameworks.

In another (more convincing) part of his paper, Quine argued that there is no way to make scientific sense of a sharp analytic/synthetic distinction. He connected this point to his holism about testing. For Quine, all our ideas and hypotheses form a single "web of belief," which has contact with experience only as whole. An unexpected observation can prompt us to make a great variety of possible changes to the web. Even sentences that might look analytic can be revised in response to experience in some circumstances. Quine noted that strange results in quantum physics had suggested to some that revisions in logic might be needed.

In this discussion of problems for logical positivism, I have included some discussions that started early and some that took place after World War II, when the movement had begun its U.S.-based transformation. I'll now turn to these later stages of the movement.

2.5 *Logical Empiricism and the Web of Belief*

Let's see how things looked in the years after World War II. Schlick is dead, and other remnants and allies of the Vienna Circle—Carnap, Hempel, Reichenbach, and Feigl—are safely housed in American universities. Many of the same people are involved, but the work is now different. The revolutionary attempt to destroy traditional philosophy has been replaced by a program of careful logical analysis of language and science. Discussion of the contributions that could be made by the scientific worldview to a democratic socialist future have been dropped or greatly muted.

(Despite this, the FBI collected a file on Carnap as a possible communist sympathizer.)

As before, ideas about language guided logical empiricist ideas about science. (Remember that I am using the term "logical empiricism" rather than "logical positivism" for this later phase.) The analytic/synthetic distinction had not been rejected, but it was regarded as questionable. The verifiability theory, which had been so scythe-like in its early forms, was replaced with a *holistic empiricist theory of meaning.* Theories were seen as structures that connect many hypotheses together. These structures are connected, as wholes, to the observable realm, but any *part* of a theory—a claim or hypothesis or concept—does not have some specific set of observations associated with it. A theoretical term, such as "electron" or "gene," derives its meaning from its place in the whole structure and from the structure's connection to the realm of observation.

Late in the logical empiricist era, in 1970, Feigl gave a pictorial representation of what he called the "orthodox view" of theories (see fig. 2.2). A network of theoretical hypotheses ("postulates") is connected by stages to what Feigl calls the "soil" of experience. This anchoring is the source of the network's meaning. Feigl used this picture to describe a single scientific theory. For the more extreme holism of Quine, a person's total set of beliefs forms a single network.

The logical positivist distinction between observational and theoretical parts of language was kept roughly intact. But the idea that observational language describes private sensations had been dropped. The observational base of science was seen as being made up of descriptions of observable physical objects (though Carnap thought it might occasionally be useful to work with a language referring to sensations).

Logical positivist views about the role of logic in philosophy, and about the sharp separation between the logic of science and the historical and psychological side of science, were basically unchanged. A good example of the kind of work done by logical empiricists is provided by their work on explanation in science (see especially Hempel and Oppenheim 1948; Hempel 1965). For Hempel, to explain something is to show how to infer it using a logical argument, where the premises of the argument include at least one statement of a natural law (see chapter 11).

We saw that logical positivism held that the sole aim of science is

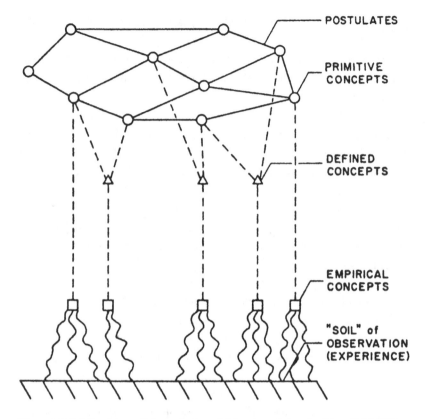

Figure 2.2. Feigl's picture of the logical empiricist view of theories (from Feigl 1970; reproduced courtesy of the University of Minnesota Press)

to track patterns in experience. For logical positivism, when a scientist seems to be trying to describe unobservable structures in the world that give rise to what we see, the scientist must instead be seen as describing the observable world in a special, abstract way. Scientific language is only meaningful insofar as it picks out patterns in the flow of experience. Does logical *empiricism*, the later version of the view, make the same claim? Does logical empiricism claim that scientific language only describes patterns in observables?

The answer is that logical empiricists agonized over this. In their hearts their answer was yes, but this answer seemed to get harder and harder to defend. Carl Hempel wrote an article in 1958 called "The Theoretician's Dilemma" that was the height of logical empiricist agony over the issue. As an empiricist, Hempel was attracted to the idea that the only possible role for those parts of language that seem to refer to unobservable entities is to help us pick out patterns in the observable realm. And if the parts of theories that appear to posit unobservable things are really any good, this "goodness" has to show up in advantages the theory has in its handling of observables. So there is no justification for seeing these parts of scientific language as describing real objects lying beyond experience. But Hempel and the logical empiricists found themselves forced to concede that this view does not make much sense of actual scientific work. When scientists use terms such as "electron" or "gene," they act as if they are doing more than tracking complex patterns in the observable realm. The idea that the logical empiricists were being pushed toward—the idea that scientific theories are aimed at describing unobservable real structures and processes—was hard to put on the table and defend, though. Empiricist philosophy of language seemed implacably opposed to it.

Empiricists were familiar with bad versions of the idea that behind the ordinary world of observables there is a special and superior realm, pure and perfect. This "layered" view of reality seemed to empiricists a source of endless trouble, right from the time of the ancient Greek philosopher Plato, who distinguished the illusory, unstable world of "appearances" from the more perfect and real world of "forms." Empiricists have rightly been determined to avoid this kind of picture. But much of science does appear to be a process in which people hypothesize hidden structures that give rise to observable phenomena. These hidden structures are not pure and perfect, or "more real" than the observable parts of the world, but they do lie behind or beneath observable phenomena. Of course, unobservable structures posited by a theory at one time might well turn out to be observable at a later time. In science, there is no telling what kinds of new access to the hidden parts of the world we might eventually achieve. But still, much of science does seem to proceed by positing entities that are, at the time of the research in question, truly hidden. For the traditional empiricist philosopher, understanding scientific theorizing in a way that posits a layer of observable phenomena and a layer of hidden structure responsible for the phenomena takes us far too close to bad old philosophical views like Plato's. We are too close for comfort, so we must give a different kind of description of how science works.

The result is an insistence that, ultimately, the only thing scientific language can do is describe patterns in the observable realm. In the first published paper that introduced logical positivism, Carnap, Hahn, and Neurath wrote, "In science there are no 'depths'; there is surface everywhere" ([1929] 1973, 306). This is a vivid expression of the empiricist aversion to a view in which the aim of theorizing is to describe hidden levels of structure. Science uses unusual theoretical concepts (which look initially like attempts to refer to hidden things) as a way of discovering and describing subtle patterns in the observable realm. So the logical positivists and the logical empiricists talked continually about *prediction* as the goal of science. Prediction was a substitute for the more obvious-looking—but ultimately forbidden—goal of describing the real hidden structure of the world.

Twentieth-century empiricism made an important mistake here. We can make sense of science only by treating much of it as an attempt to describe hidden structures that give rise to observable phenomena. This is a version of *scientific realism*, an idea that will be discussed later in this book. In science there *are* depths. There is not a simple and fixed distinction between two layers in nature—the empiricists were right to distrust this idea. Instead, there are many layers, or rather a continuum between structures that are more accessible to us and structures that are less accessible. Genes are hidden from us in some ways, but not as hidden as protons, which in turn are not as hidden as quarks. Although there are depths in science, what is deep at one time can come to the surface at later times, and there may be lots of ways of interacting with what is presently deep.

Some people associated with the logical empiricist movement did try to break away from their assumptions about what can be meaningfully spoken about. Herbert Feigl was part of the movement from the early years, and he came to think that a mistake had been made here. Feigl, too, used metaphors of depths and surfaces to describe what was going on. He said the logical positivists were right to emphasize the role of observation in providing evidence for our claims, but they pushed this emphasis so far that they lost sight of the objects that these claims are typically about: "Dazzled by the admittedly tremendous importance of the evidential basis for our knowledge claims, positivists have regrettably neglected the very objects of those knowledge claims. They have myopically flattened them into the surface of evidence" (1958, 98). The word "myopically," whose literal meaning refers to nearsightedness, can be used in a metaphorical way to mean that things are not being seen clearly. But Feigl, cleverly, seems to use the word with something closer to its literal meaning. The positivists, with their eyes fixed on evidence, were not seeing what lies beyond that evidence.

I think that quite a few empiricists in the middle of the twentieth century would have liked to decisively abandon the idea that all we can ultimately do in science is describe patterns in our experience, or our observational evidence. But this would have meant going completely against the empiricist ideas about meaning and testing that had energized the movement and had seemed such an advance in the early years. And it seemed very hard to come up with a better view about how scientific language works and what kind of meaning a theory has. This still seems hard today.

2.6 *Experience, Experiment, and Action*

Logical empiricist ideas dominated much of American philosophy and were very influential elsewhere in the English-speaking world and in some parts of Europe in the middle of the twentieth century. But by the mid-1960s the view was definitely under threat, and by the middle or late 1970s logical empiricism was close to extinction. How the decline occurred is described in the next few chapters.

At the end of this chapter on empiricism, though, I want to introduce one more theme that will come up in several contexts in this book. I'll do so with a quote from the twentieth-century physicist Richard Feynman.

The principle of science, the definition, almost, is the following: *The test of all knowledge is experiment.* Experiment is the *sole judge* of scientific "truth."

This is from *The Feynman Lectures on Physics*, a text written for a famous course he taught in the 1960s. In some ways, what Feynman says is similar to what empiricist philosophers like to say, but in one way it looks different. Experiment, he says, is what matters. Not observation or experience, but experiment. And he says that experiment is the sole judge in science.

This is a surprising thing to say. It seems that some observations come from experiments and some do not, and those that do not have sometimes been very important in science. Astronomy is a field where careful observation has been done for many centuries, but not a lot of experiment. And astronomy was perhaps the first of all sciences to reach an advanced state, making impressive predictions by the seventeenth century and certainly the eighteenth. In 1705 Edmund Halley used Newton's theory, along with past records, to predict that a comet would return in 1758. It did exactly that (though it was a close call, arriving on Christmas Day). This was rightly seen as a triumph. It would be hard to view early modern astronomy as a deficient science. (Darwin's work on the theory of evolution, and evolutionary biology for many years after, was also not greatly based on experiment, though Darwin did experiments in other areas. "Experimental evolution" is a fairly new subfield of biology, though an important one.)

In the case of astronomy, you might reply that there have, in fact, been experiments for a long time. We often think of an experiment as setting up an arrangement to see what will happen (when I add this chemical to this other chemical, and so on). But perhaps what is basic to experiment is just having an organized plan where you will do *something*—go through some procedure—and observe the results in a particular way. This might include pointing a telescope at the sky every night at a particular time.

Alternatively, Feynman might think that any science has to head toward experiment as much as possible, even if some progress is possible without it. That is a possibility to consider, though it seems different from what Feynman said.

This whole issue—the relation between observation and experiment—was for a long time largely ignored by empiricist philosophers. In some ways, this is symptomatic of something even bigger. The role of action, and the ways we deliberately transform the world around us by doing things, building things, and so on, was also neglected, not just in philosophical writing about science but in much empiricist philosophy generally. This is not only true of the old figures from many centuries ago; it is also true of Quine, the empiricist who criticized the dogmas of others and defended his web-of-belief view. When Quine talks about what we use our web of belief for, he emphasizes prediction—anticipating what will happen. He hardly ever talks about using our beliefs to work out how to act.

Action is important here in at least two ways. First, much of the point of learning about the world is working out what to do—what material to build bridges out of, how to purify water for drinking. Also, when you act you often change your own experience. This is true both in everyday cases—when you unwrap a box, walk down the street, or turn on the lights—and in more complicated cases, when we build a particle accelerator that can create observations that would never otherwise be possible. There are many questions to think about here. What distinguishes experiment from other activities? The important issue is not so much to work out what counts as "experiment," in the usual sense of this term, but to work out which, if any, of the more active forms of observation make a difference to knowledge and investigation. Is it always good for a science to head to down the road toward experiment, or experiment of some particular kind? If so, why?

Further Reading and Notes

Schlick's "Positivism and Realism" (1932–33) and Feigl's "Logical Empiricism" (1943) are good statements of logical positivism by original members of the Vienna Circle. (Feigl uses the term "logical empiricism," but his paper describes a fairly undiluted version of the view.) Ayer's

Language, Truth, and Logic (1936) is readable, vivid, and exciting. Some see it as a distortion of logical positivist ideas. Reichenbach's *The Rise of Scientific Philosophy* (1951), which I discuss in section 2.2, is very good although Reichenbach was not part of the Vienna Circle itself.

Classics of the empiricist tradition include Locke's *An Essay Concerning Human Understanding* (1690), Berkeley's *Three Dialogues between Hylas and Philonous* (1713), and Hume's *An Enquiry Concerning Human Understanding* (1748 [1999]), all of which are available in free online editions. Reichenbach's Hegel quote is from *The Philosophy of History* ([1824] 1956). Neurath's boat metaphor ("sailors who have to rebuild their ship on the open sea") appears in his "Anti-Spengler" (1921). Neurath is discussed in detail in Cartwright et al. (1996). There are many collections of articles about the logical positivists (/empiricists) now, including Giere and Richardson (1996). Galison's "Aufbau/Bauhaus" (1990) is a fascinating account of the artistic, social, and political interests of the logical positivists and the links between these interests and their philosophical ideas.

The logical positivists were inspired by Einstein, and in some ways he repaid the compliment. His paper "On the Method of Theoretical Physics" (1934) expresses ideas quite close to theirs.

Hempel's *Aspects of Scientific Explanation* (1965) is the most influential statement of the later, moderate "logical empiricist" view. His *Philosophy of Natural Science* (1966) is the easy version. Carnap's later lectures have been published as *Introduction to the Philosophy of Science* (1995). The Feynman quote is from the first page of Feynman, Leighton, and Sands (1963–65), chapter 1.

Chapter 3

Evidence
and Induction

3.1 *The Mother of All Problems*

In this chapter we begin looking at a very important and difficult problem, the problem of understanding how observations can provide evidence for a scientific theory. In some ways, this has been *the* fundamental problem over the last hundred years of philosophy of science. This problem was central to the project of logical empiricism, and it was a source of constant frustration. Abandoning logical empiricism does not make the problem go away, though. In some form or other, it arises for nearly everyone.

The aim of the logical empiricists was to develop a logical theory of evidence in science. In this chapter I'll often use the terminology they favored, in which the *confirmation* of theories is what has to be understood. Confirmation is not the same as proof; a theory can be confirmed by evidence but turn out to be wrong. Confirmation is a kind of support that theories can get from evidence, where this support is usually partial rather than decisive. The logical empiricists wanted to treat confirmation as an abstract relation between sentences. It has become fairly clear that their approach to the problem is doomed. The way to analyze evidence in science is to develop a different kind of theory. But it will take a lot of discussion, in this and later chapters, before the differences between approaches that will and will not work can emerge. The present chapter will look mostly at how the problem of evidence was tackled in the middle of the twentieth century. And that is a tale of woe.

Before looking at twentieth-century work on these issues, we must again spend some time further back in the past. The confirmation of theories is closely connected to another classic topic in philosophy: the *problem of induction*. What reason do we have for expecting patterns observed in our past experience to hold in the future? What justification do we have for using past observations as a basis for generalization about things we have not yet observed? The most famous discussions of induction come from the eighteenth-century Scottish empiricist David Hume. Hume asked, what reason do we have for thinking that the future will resemble the past? There is no contradiction in supposing that the future could be totally unlike the past. It is possible that the world could change radically at any point, rendering previous experience useless. How do we know this will not happen? We might say to Hume that when we have relied on past experience before, it has turned out well for us. The policy has a good track record. But Hume will reply that this is begging the question—presupposing what has to be shown. Induction has worked in the past, yes, but that's the *past*. We have successfully used "past pasts" to tell us about "past futures." But our problem is whether anything about the past gives us good information about what will happen tomorrow.

Hume concluded that we have no reason to expect the past to resemble the future. Hume was an "inductive skeptic." He accepted that we all use induction to make our way around the world. And he was not suggesting that we stop doing so (even if we could). Induction is psychologically natural to us. Despite this, Hume thought it had no rational basis; it is just a habit we all have. Hume's inductive skepticism has haunted empiricism ever since. The problem of confirmation is not the same as the classical problem of induction, but the two are closely related.

3.2 *Induction, Deduction, Confirmation, and Explanatory Inference*

The logical empiricists tried to show how observational evidence could provide support for a scientific theory. Again, there was no attempt to show that scientific theories can be proved. Error is always possible, but evidence can support one theory over another. The cases that were to be covered by this analysis included the simplest and most traditional cases of induction: if we see a multitude of cases of white swans, and no other colors, why does that give us reason to believe that all swans are white? But obviously not all cases of evidence in science are like this.

The observational support for Copernicus's theory that the Earth goes around the sun, and for Darwin's theory of evolution, seem to work very differently from a traditional induction. Darwin did not observe a set of individual cases of evolution and then generalize.

The logical empiricists wanted a theory of evidence that would cover all these cases. They were not trying to develop a recipe for confirming theories. Rather, the aim was to give an account of the relationships between the statements that make up a scientific theory and statements describing observations, which make the observations support the theory.

Next, I should say more about the distinction between deductive and inductive logic (a distinction introduced in chapter 2). Deductive logic is the well-understood and less controversial kind of logic. It is a theory of patterns of argument that transmit truth with certainty. These arguments have the feature that if the premises of the argument are true, the conclusion is guaranteed to be true. An argument of this kind is *deductively valid*. The most famous example of a logical argument is a deductively valid argument:

premises   All men are mortal.
                 Socrates is a man.

                 _____

conclusion   Socrates is mortal.

A deductively valid argument might have false premises. In that case, the conclusion might be false as well (although it also might not be). What you get out of a deductive argument depends on what you put into it.

The logical empiricists thought that deductive logic could not serve as a complete analysis of evidence and argument in science. Scientific theories do have to be logically consistent, but this is not the whole story. Many inferences in science are not deductively valid, but they still can be good inferences; they can provide support for their conclusions.

For the logical empiricists, there is a reason why so much inference in science is not deductive. As empiricists, they believed that all our evidence derives from observation. Observations are always of *particular* objects and occurrences. But the logical empiricists thought that the great aim of science is to discover and establish *generalizations*. Sometimes the aim was seen as describing "laws of nature," though this concept was also regarded with some suspicion (see chapter 11). The central idea was that science aims at formulating and testing generalizations, and these generalizations were seen as having, at least potentially, an infinite range of application. No finite number of observations can conclusively establish a generalization of this kind, so inferences from observations in support of generalizations are always nondeductive. In contrast, all it takes is one case of the right kind to prove a generalization to be *false*; this fact will loom large in the next chapter.

In many discussions of these topics, the logical empiricists (and some later writers) used a simple terminology in which all arguments are either deductive or inductive. Inductive logic was thought of as a theory of all good arguments that are not deductive. But this terminology can be misleading, and I will set things up differently.

I will use the term "induction" only for inferences that go from particular observations to generalizations. To use the most traditional example, the observation of a large number of white swans (and no swans of any other color) might be used to support the hypothesis that all swans are white. We could express the premises with a list of particular cases: "Swan 1 observed at time t_1 was white; swan 2 observed at time t_2 was white...." Or we might simply say: "All the many swans observed so far have been white." The conclusion will be the claim that all swans are white—a conclusion that could well be false but is supported, to some extent, by the evidence. Sometimes the terms "enumerative induction" or "simple induction" are used for inductive arguments of this most traditional kind. Not all inferences from observations to generalizations have this form, though. (And a note to mathematicians: mathematical induction is really a kind of deduction, even though it has the superficial form of induction.)

A form of inference closely related to induction is *projection*. In a projection, we infer from a number of observed cases to arrive at a prediction about the next case, not to a generalization about all cases. We see a number of white swans and infer that the next swan will be white. Obviously there is a close relationship between induction and projection, though there are a variety of ways of understanding this relationship.

Other kinds of nondeductive inference are seen often in science and everyday life. For example, during the 1980s Luis and Walter Alvarez began claiming that a huge meteor had hit the Earth about 65 million years ago, causing a massive explosion and dramatic weather changes that coincided with the extinction of the dinosaurs (Alvarez et al. 1980). The Alvarez team claimed that the meteor caused the extinctions, but let's leave that aside here. Consider just the hypothesis that a huge meteor hit the Earth 65 million years ago. A key piece of evidence for this hypothesis is the presence of unusually high levels of some rare chemical elements, such as iridium, in layers in the Earth's crust that are 65 million years old. These elements tend to be found in meteors in much higher concentrations than they are near the surface of the Earth. This observation was taken to be good evidence supporting the theory that a meteor hit the Earth around that time.

If we set this case up as an argument, with premises and a conclusion, it clearly is not an induction or a projection. We are not inferring to a generalization, but to a hypothesis about a structure or event that would explain our data. Several terms are used in philosophy for inferences of this kind. C. S. Peirce called these "abductive" inferences as opposed to inductive ones. Others have called them "explanatory inductions" or "theoretical inductions." The most common term for them is "inference to the best explanation," but I will use a slightly different one— "explanatory inference."

So I will recognize two main kinds of nondeductive inference, induction and explanatory inference (along with projection, which is closely linked to induction). The problem of analyzing evidence includes all of these.

How are these kinds of inference related to each other? For many logical empiricists and others, induction is the most fundamental kind of nondeductive inference. Reichenbach claimed that all nondeductive inference in science can be reconstructed in a way that depends only on a form of inference that is close to traditional induction. What looks like an explanatory inference can be somehow broken down and reconstructed as a complicated network of inductions and deductions. Carnap did not make this claim, but he did seem to view induction as a model for all other kinds of nondeductive inference. Understanding induction was in some sense the key to the whole problem.

So one way to view the situation is to see induction as fundamental.

But it is also possible to do the opposite, to claim that explanatory inference is fundamental. Gilbert Harman argued in 1965 that inductions are justified only when they are explanatory inferences in disguise, and others have followed up this idea in various ways.

Explanatory inference seems much more common in science than induction. In fact, you might be wondering whether science contains any inductions of the simple, traditional kind. That suspicion is reasonable, but it turns out that science does contain inferences, including important ones, that at least look like traditional inductions. Here is one example. During the work that led to the discovery of the structure of DNA by James Watson and Francis Crick, a key piece of evidence was provided by "Chargaff's rules." These rules, described by Erwin Chargaff in the late 1940s, have to do with the relation between the amounts of the four bases—C, A, T, and G—that help make up DNA. Chargaff found that in the DNA samples he analyzed, the amounts of C and G were always roughly the same, and the amounts of T and A were always roughly the same (Chargaff 1951). This fact about DNA became important in the discussions of how DNA molecules are put together. I called it a "fact" just above, but of course Chargaff had not observed *all* the molecules of DNA that exist, and neither have we. Back then, Chargaff's claim rested on an induction from a small number of cases (though he did cover a diverse range of organisms). Today we can give an argument for why Chargaff's rules hold that is not just a simple induction; the structure of DNA explains why Chargaff's rules must hold. But it might seem that, back when the rules were originally discovered, the only reason to take the rules to describe all DNA was inductive.

It probably seems a good idea, then, to refuse to treat one of these kinds of inference as more fundamental than the other. Perhaps there is more than one kind of good nondeductive inference (and maybe there are others besides the ones I have mentioned). Philosophers often find it attractive to think that there is ultimately just one kind of nondeductive inference, because that seems to be a simpler situation. But this argument from simplicity is not very convincing.

Let us return to our discussion of how the problem was handled by the logical empiricists. They used two main approaches. One was to formulate an inductive logic that looked as much as possible like deductive logic. That was Hempel's approach. The other approach, used by Carnap, was to apply the mathematical theory of probability. In the next two sections of this chapter, I will discuss some famous problems for logical empiricist theories of confirmation. The problems are especially easy to discuss in the context of Hempel's approach. An examination of Carnap's view is beyond the scope of this book. Over the course of his career, Carnap developed very sophisticated models of confirmation using probability theory applied to artificial languages. Problems kept arising. More and more special assumptions were needed to make the results come out in a way that looked reasonable. There was never a knockdown argument against him, but the project came to seem less and less relevant to science, and it eventually ran out of steam.

Although Carnap's approach to analyzing confirmation did not work out, the idea of using probability theory to understand confirmation remains popular and has been developed in new ways. Certainly this looks like a good approach; it does seem that observing the raised iridium level in the Earth's crust made the Alvarez meteor hypothesis *more probable* than before. In chapter 12 I will describe new ways of using probability theory to understand the confirmation of theories.

Before moving on to some famous puzzles, I will discuss a simple proposal that may have occurred to you.

The term "hypothetico-deductivism" is used in several ways by people writing about science. Sometimes it is used to describe a view about testing and confirmation. According to this view, hypotheses in science are confirmed when their logical consequences turn out to be true. This idea covers a variety of cases; the confirmation of a white-swan generalization by observing white swans is one case, and another is the confirmation of a hypothesis about an asteroid impact by observations of the true consequences of this hypothesis.

As Clark Glymour has emphasized (1980), an interesting thing about this idea is that it is hopeless when expressed in a simple way, but something like it seems to fit well with many episodes in the history of science. One problem is that a scientific hypothesis will only have consequences of a testable kind when it is combined with other assumptions, as we have seen. But set that problem aside for a moment. The suggestion above is that a theory is confirmed when a true statement about observables can be derived from it.

This claim is vulnerable to many objections. For example, any theory *T* deductively implies *T-or-S*, where *S* is any sentence at all. But *T-or-S* can be conclusively established by observing the truth of *S*. Suppose *S* is observational. Then we can establish the truth of *T-or-S* by observation, and that confirms *T*. This is obviously absurd. Similarly, if theory *T* implies observation *E*, then the theory *T&S* implies *E* as well. So *T&S* is confirmed by *E*, and *S* here could be anything at all. (Note the similarity here to a problem discussed at the beginning of section 2.4.) There are many more cases like this.

The situation is indeed strange. People often regard a scientific hypothesis as supported when its consequences turn out to be true; this is taken to be a routine and reasonable part of science. But when we try to summarize this idea using simple logic, it seems to fall apart. Does the fault lie with the original idea, with our summary of the idea using basic logic, or with basic logic itself?

The logical empiricists' response, here and elsewhere, was to hang steadfastly onto the logic, and often to hang onto their preferred ways of translating scientific theories into a logical framework as well. This led them to question some quite reasonable-looking ideas about evidence and testing. It also led to a situation in which philosophy of science seemed to turn into an exercise in "logic-chopping" for its own sake. And as we will see in a moment, even the logic-chopping did not go well. Despite all this, there is a lot to learn from the problems faced by logical empiricism. Confirmation really is a puzzling thing. Let us look at some famous puzzles.

3.3 *The Ravens Problem*

The logical empiricists put a great deal of work into analyzing the confirmation of generalizations by observations of their instances. At this point we will switch birds, in accordance with tradition. How is it that repeated observations of black ravens can confirm the generalization that all ravens are black?

First I will deal with a simple suggestion that will not work. Some readers might be thinking that if we observe a large number of black ravens and no nonblack ones, then at least we are cutting down the number of ways in which the hypothesis that all ravens are black might be wrong. With each raven we see, there is one fewer raven that might fail to fit the theory. In some sense, the chance that the hypothesis is true should be slowly increasing. But this does not help much. First, the logical empiricists were concerned to deal with the case where generalizations cover an infinite number of instances. In that case, as we see each raven we are not reducing the number of ways in which the hypothesis might fail. Also, note that even if we forget this problem and consider a generalization covering just a finite number of cases, the kind of support that is analyzed here is a weak one. That is clear from the fact that we get no help with the problem of projection. As we see each raven, we know there are fewer and fewer ways for the generalization to be false, but this does not tell us anything about what to expect with the next raven we see.

So let us look at the problem differently. Hempel suggested that, as a matter of logic, all observations of black ravens confirm the generalization that all ravens are black. More generally, any observation of an *F* that is also *G* supports the generalization "All *F*s are *G*." He saw this as a basic fact about the logic of support.

This looks like a reasonable place to start. And here is another obvious-looking point: any evidence that confirms a hypothesis *H* also confirms any hypothesis that is logically equivalent to *H*.

What is logical equivalence? Think of it as what we have when two sentences say the same thing in different terms. More precisely, if *H* is logically equivalent to *H**, then it is impossible for *H* to be true but *H** false, or vice versa.

But these two innocent-looking claims generate a problem. In basic logic, the hypothesis "All ravens are black" is logically equivalent to "All nonblack things are not ravens." Let us look at this new generalization. "All nonblack things are not ravens" seems to be confirmed by the observation of a white shoe. The shoe is not black, and it's not a raven, so it fits the hypothesis. But given the logical equivalence of the two hypotheses, anything that confirms one confirms the other. So the observation of a white shoe confirms the hypothesis that all ravens are black! That sounds ridiculous. As Nelson Goodman (1955) put it, we seem to have the chance to do a lot of "indoor ornithology"; we can investigate the color of ravens without ever going outside to look at one.

This simple-looking problem is hard to solve. Debate about it continues to this day. Hempel himself was well aware of this problem—he was the one who originally thought of it. But there has not been a solution proposed that everyone, or even a majority of people, has agreed upon.

One possible reaction is to accept the conclusion. This was Hempel's response. Observing a white shoe does confirm the hypothesis that all ravens are black, though presumably only by a tiny amount. Then we can keep our simple rule that whenever we have an "All *F*s are *G*" hypothesis, any observation of an *F* that is *G* confirms it and also confirms everything logically equivalent to "All *F*s are *G*." Hempel stressed that, logically speaking, an "All *F*s are *G*" statement is not a statement about *F*s but a statement about everything in the universe—the statement that if something is *F*, then it is *G*. We should note that according to this reply, the observation of the white shoe also confirms the hypothesis that all ravens are green, that all aardvarks are blue, and so on. Hempel was comfortable with this situation, but most others have not been.

A multitude of other solutions have been proposed. I will discuss just two ideas, which I regard as being on the right track.

Here is the first idea. Perhaps observing a white shoe or a black raven may or may not confirm "All ravens are black." It depends on other factors. Suppose we know, for some reason, that either (1) all ravens are black and ravens are extremely rare, or else (2) most ravens are black, a few are white, and ravens are common. Then a casual observation of a black raven will support (2), a hypothesis that says that not all ravens are black. If all ravens were black, we should not be seeing them at all. Similarly, observing a white shoe may or may not confirm a hypothesis, depending on what else we know. This reply was first suggested by I. J. Good (1967).

Good's move is very reasonable. We see here a connection to the issue of holism about testing, discussed in chapter 2. The relevance of an observation to a hypothesis is not a simple matter of the content of the two statements; it depends on other assumptions as well. This is so even in the simple case of a hypothesis like "All *Fs* are *G*" and an observation like "Object *A* is both *F* and *G*." Good's point also reminds us how artificially simplified the standard logical empiricist examples are. No biologist would seriously wonder whether seeing thousands of black ravens makes it likely that all ravens are black. Our knowledge of genetics and bird coloration leads us to expect some variation, such as cases of albinism, even when we have seen thousands of black ravens and no other colors.

Here is a second suggestion about the ravens, which is consistent with Good's idea but goes further. Whether or not a black raven or a white shoe confirms "All ravens are black" might depend on the order in which you learn of the two properties of the object.

Suppose you hypothesize that all ravens are black, and someone comes up to you and says, "I have a raven behind my back; want to see what color it is?" You should say yes, because if the person pulls out a white raven, your theory is refuted. You need to find out what is behind his back. But suppose the person comes up and says, "I have a black object behind my back; want to see whether it's a raven?" Then it does not matter to you what is behind his back. You think that all ravens are black, but you don't have to think that all black things are ravens. In both cases, suppose the object behind his back is a black raven and he does show it to you. In the first situation, your observation of the raven seems relevant to your investigation of raven color, but in the other case it's irrelevant.

So perhaps the "All ravens are black" hypothesis is only confirmed by a black raven when this observation had the potential to refute the hypothesis, only when the observation was part of a genuine test.

Now we can see what to do with the white shoe. You believe that all ravens are black, and someone comes up and says, "I have a white object behind my back; want to see what it is?" You should say yes, because if he has a raven behind his back your hypothesis is refuted. He pulls out a shoe, however, so your hypothesis is OK. Then someone comes up and says, "I have a shoe behind my back; want to see what color it is?" In this case you need not care. It seems that in the first of these two cases, you have gained some support for the hypothesis that all ravens are black. In the second case, you have not.

So perhaps some white-shoe observations do confirm "All ravens are black," and some black-raven observations don't. Perhaps there is only confirmation when the observations arise during a genuine test, a test that has the potential to disconfirm as well as confirm.

Hempel saw the possibility of a view like this. His responses to Good's argument and to the order-of-observation point were similar, in fact. He said he wanted to analyze a relation of confirmation that exists just between a hypothesis and an observation itself, regardless of extra information we might have, and regardless of the order in which observations were made. But perhaps Hempel was wrong; perhaps there is no such relation. We cannot answer the question of whether an observation of a black raven confirms the generalization unless we know something about the way the observation was made and unless we make assumptions about other matters as well.

Hempel thought that some observations are just "automatically" relevant to hypotheses, regardless of what else is going on. That is true in the case of the deductive refutation of generalizations; no matter how we come to see a nonblack raven, that is bad news for the "All ravens are black" hypothesis. But what is true for deductive disconfirmation is not true for confirmation.

Clearly this discussion of order-of-observation does not entirely solve the ravens problem. Why does order matter? And what if both properties are observed at once? I will return to this issue in chapter 12, using a more complex framework. Putting it briefly, perhaps we can only understand confirmation and evidence by taking into account the procedures involved in generating data.

I will make one more comment on the ravens problem. This one is a digression, but it does help illustrate what is going on. In psychology there is a famous experiment called the "selection task" (Wason and Johnson-Laird 1972). The experiment has been used to show that many people (including highly educated people) make bad logical errors in certain circumstances. The experimental subject is shown four cards with half of each card masked. The subject is asked to answer this question: "Which masks do you have to remove to know whether it is true that if

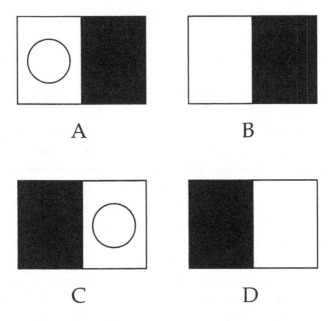

Figure 3.1. The Wason selection task

there is a circle on the left of a card, there is a circle on the right as well?" Have a look at figure 3.1 and try to answer the question yourself before reading the next paragraph.

Large majorities of people in many (though not all) versions of this experiment give the wrong answer. Many people tend to answer "only card *A*" or "card *A* and card *C*." The right answer is *A* and *D*. Compare this to the ravens problem; the problems have the same structure. I am sure Hempel would have given the right answer if he had been a subject in the four-card experiment, but the selection task might show something interesting about why confirmation has been hard to analyze. For some reason it is difficult for people to see the importance of "card *D*" tests in cases like this, and it is easy for people to think, wrongly, that "card *C*" tests are important. If you are investigating the hypothesis that all ravens are black, card *D* is analogous to the situation when someone says he has a white object behind his back. Card *C* is analogous to the situation where he says he has a black object behind his back. Card *D* is a real test of the hypothesis, but card *C* is not. Unmasking card *C* is evidentially useless, even though it may fit with what the hypothesis says. Not all observations of cases that fit a hypothesis are relevant as tests.

3.4 Goodman's "New Riddle of Induction"

In this section I will describe an even more famous problem, revealed by Nelson Goodman (1955). This argument looks strange, and it is easy to misinterpret. But the issues it raises are very deep.

Goodman's immediate goal was to show that there cannot be a purely "formal" theory of confirmation. He does not think that confirmation is impossible, or that induction is a myth. He just thinks they work differently from the way many philosophers have thought.

What would a formal theory of confirmation look like? The easiest way to explain this is to look at deductive arguments. Recall the most famous deductively valid argument:

Argument 1

premises   All men are mortal.

Socrates is a man.

conclusion   Socrates is mortal.

The premises, if they are true, guarantee the truth of the conclusion. But the fact that the argument is a good one does not have anything in particular to do with Socrates or manhood. Any argument that has the same form is just as good. That form is as follows:

All *F*s are *G*.
Object *a* is *F*.

Object *a* is *G*.

Any argument with this form is deductively valid, no matter what we substitute for "*F*," "*G*," and "*a*." As long as the terms we put in do pick out definite properties or classes of objects, and as long as the terms retain the same meaning all the way through the argument, the argument will be valid.

The deductive validity of arguments depends only on the form or pattern of the argument, not the content. This is something the logical empiricists wanted to build into their theory of induction and confirmation. Goodman aimed to show that this is impossible; there can never be a formal theory of induction and confirmation.

How did Goodman do it? Consider argument 2.

Argument 2
All the many emeralds observed, in diverse circumstances, prior to the year 2050 have been green.

All emeralds are green.

This looks like a good inductive argument. (Like some of the logical empiricists, I use a double line between premises and conclusion to indicate that the argument is not supposed to be deductively valid.) The argument does not give us a guarantee; inductions never do. And if you would prefer to express the conclusion as "Probably, all emeralds are green," or something similar, that will not make a difference to the rest of the discussion. (If you know something about minerals, you might object that emeralds are often regarded as green by definition: emeralds are beryl crystals made green by trace amounts of chromium. Please just regard this as another unfortunate choice of example by the literature.)

Now consider argument 3:

Argument 3
All the many emeralds observed, in diverse circumstances, prior to the year 2050 have been grue.

All emeralds are grue.

Argument 3 uses a new word, "grue." We define "grue" as follows:

GRUE: An object is *grue* if and only if it was first observed before the year 2050 and is green, *or* if it was not first observed before 2050 and is blue.

The world contains lots of grue things; there is nothing strange about grue objects, even though there is something strange about the word. The grass outside my door as I write this is grue. The sky outside on July 1, 2055, will be grue, if it is a clear day. An individual object does not have to change color in order to be grue—this is a common misinterpretation. Anything green that is at some point observed before 2050 passes the test for being grue. So, all the emeralds we have seen so far have been grue.[1]

Argument 3 does not look like a good inductive argument. Argument 3 leads us to believe that emeralds observed in the future will be blue, on the basis of previously observed emeralds being green. The argument also conflicts with argument 2, which looks like a good argument. But arguments 2 and 3 have exactly the same form. That form is as follows:

All the many *E*s observed, in diverse circumstances, prior to 2050 have been *G*.

All *E*s are *G*.

We could represent the form even more schematically than this, but that does not matter to the point. Goodman's point is that two inductive arguments can have the exact same form, but one argument can be good

---

1. You are probably reading this before 2050, and might then wonder how we know the premise to be true. Why not say an object is grue if and only if it has been observed before now and is green, or has never been observed and is blue? It is fine to present things that way, as long as we treat "now" in the argument as a fixed time, not something that keeps moving. I think it's easier to get a grip on the argument by choosing a particular date in the future and supposing that all the emeralds seen before that date are green.

while the other is bad. So what makes an inductive argument a good or bad one cannot be just its form. Consequently, there can be no purely formal theory of induction and confirmation. Note that the word "grue" works perfectly well in *de*ductive arguments. You can use it in the form of argument 1, and it will cause no problems. But induction is different.

Suppose Goodman is right so far. We then need to work out what is wrong with argument 3. This is the new riddle of induction. The obvious thing to say is that there is something about the word "grue" that makes it inappropriate for use in inductions. A good theory of induction should include a restriction of some kind on the terms that occur in inductive arguments. "Green" is OK and "grue" is not.

This has been the most common response to the problem. But as Goodman says, it is hard to spell out the details of such a restriction. Suppose we say that the problem with "grue" is that its definition includes a reference to a specific time. Goodman's reply is that whether or not a term is defined in this way depends on which language we take as our starting point. To see this, let us define a new term, "bleen."

> BLEEN: An object is *bleen* if and only if it was first observed before the year 2050 and is blue, *or* if it was not first observed before 2050 and is green.

We can use the English words "green" and "blue" to define "grue" and "bleen," and if we do so we must build a reference to time into the definitions. But suppose we spoke a language that was like English except that "grue" and "bleen" were basic, familiar terms and "green" and "blue" were not. Then if we wanted to define "green" and "blue," we would need a reference to time.

> GREEN: An object is *green* if and only if it was first observed before the year 2050 and is grue, *or* if it was not first observed before 2050 and is bleen.

(You can see how it will work for "blue.") Goodman claimed that whether or not a term "contains a reference to time" or "is defined in terms of time" is a *language-relative* matter. Terms that look OK from the standpoint of one language will look odd from another. So if we want to rule out "grue" from inductions because of its reference to time, then whether an induction is good or bad will depend on which language we treat as our starting point. Goodman thought this conclusion was fine. A good induction, for Goodman, must use terms that have a history of normal use in our community. That was his own solution to his problem. Most other philosophers did not like this at all. It seemed to say that the value of inductive arguments depended on irrelevant facts about which language we happen to use.

Consequently, many philosophers have tried to focus not on the words "green" and "grue" but on the *properties* that these words pick out, or the *classes* or *kinds* of objects that are grouped by these words. We might argue that greenness is a natural and objective feature of the world, and grueness is not. Putting it another way, the green objects make up a "natural kind," a kind unified by real similarity, while the grue objects are an artificial or arbitrary collection. Then we might say: a good induction has to use terms that we have reason to believe pick out natural kinds. Taking this approach plunges us into hard problems in other parts of philosophy. What is a property? What is a "natural kind"? These problems have been controversial since the time of Plato. (I'll have a look at them in chapter 10.)

Although Goodman's "riddle" is abstract and uses invented terms, it has interesting links to real problems in science. I think Goodman's problem encapsulates within it several distinct methodological issues. First, there is a connection between Goodman's problem and the "curve-fitting problem" in data analysis. Suppose you have a set of data points in the form of x and y values, and you want to discern a general relationship expressed by the points by fitting a function to them. The points in figure 3.2 fall almost exactly on a straight line, and that seems to give us a natural prediction for the y value we expect for $x = 4$. However, there is an infinite number of different mathematical functions that fit our three data points (as well or better) but which make different predictions for the case of $x = 4$. How do we know which function to use? Fitting a strange function to the points seems to be like preferring a grue induc-

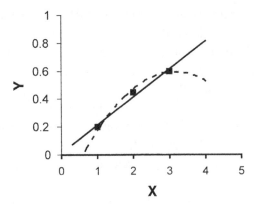

Figure 3.2. The curve-fitting problem

tion over a green induction when inferring from the emeralds we have seen.

Scientists dealing with a curve-fitting problem like this may have extra information telling them what sort of function is likely in a particular case, or they may prefer a straight line on the basis of *simplicity*. That suggests a way in which we might deal with Goodman's original problem. Perhaps the green induction is to be preferred on the basis of its simplicity?

That might work, but is it really so clear that the green-emerald induction is simpler? Goodman will argue that the simplicity of an inductive argument depends on which language we assume as our starting point, for the kinds of reasons given earlier in this section. For Goodman, what counts as a simple pattern depends on which language you speak or which categorization you assume. Hume, in the original problem of induction, wondered whether we had reason to believe that the future would resemble the past. Goodman, using his "new riddle," argued that whether or not some imagined future "resembles the past" depends on which language we habitually use.

A separate question is whether simpler theories really deserve some sort of preference in science. It is common to think so, but the issue is not so clear—this I will tackle in chapter 13.

That (almost) concludes our first foray into the problems of induction and confirmation. These problems are simple but very resistant to solution. For a good part of the twentieth century, it seemed that even the most innocent-looking principles about induction and confirmation led straight into trouble.

Some of these problems with evidence might seem to be consequences of a particular way of setting things up—an approach to evidence that puts induction in too central a place. That is roughly what I think. But here, near the end of the chapter, I want to emphasize another side to the role of induction in science, a side that involves its historical role, going back even before Hume's skeptical arguments.

In the history of science and philosophy, the idea of induction is associated with Francis Bacon, who wrote in the early seventeenth century and strongly advocated the use of induction in science. Bacon's sense of induction was not quite the same as the sense seen in logical empiricism, or Hume, but it is related. For Bacon, what was important was a contrast between different methods—different ways of approaching investigation. When Bacon wrote, he was trying first to urge people to move away from trusting ancient authorities. But once we are doing our own investigations, he wanted people to also move away from an approach in which the researcher looks for very general and basic principles first, and then seeks to understand particular things in terms of those principles. Instead, Bacon advocated a method in which you build up slowly, from lots of particular bits and pieces of data, toward bigger and more general ideas.

What Bacon had in mind in this building-up process was not just the tallying of cases (one black raven, another black raven . . .). His approach included looking for negative cases, and he also emphasized intervention in nature—experiment, not just observation. He said a scientist should "twist the lion's tail," not merely take in what's going on. All this was quite early in the Scientific Revolution, and what the idea of induction carried with it at that time was partly an ideal of *flexibility* in intellectual life. This embrace of flexibility is a very important idea, and a valuable part of the empiricist tradition. Philosophers, though, writing afterward and looking for a more systematic and exact treatment of induction, had a tendency to bring everything down to the relationship between partic-ular instances and generalizations that cover them. They have thought that the relationship between individual cases and generalizations must be the heart of the matter. Or perhaps some of them thought that this relationship was at least the right place to start—the best route in to an understanding of evidence. Neither view turned out to be right.

3.5 *Optional Section: A Little More about Hypothetico-Deductivism*

Having explored induction, it would be good to have another look at the approach to evidence I moved over quickly in an earlier part of this chapter, the *hypothetico-deductive* (HD) approach. This short section goes down some twisting paths and can be skipped.

According to a simple HD view, hypotheses in science are confirmed when their observational consequences turn out to be true. As I said, this approach initially seems closer to most science than the approach which tries to reduce everything to inductive arguments from instances to generalizations. But it also seems to run into problems as soon as you fill the idea out.

I said earlier: "One problem is that a scientific hypothesis will only have consequences of a testable kind when it is combined with other assumptions, as we have seen. But set that problem aside for a moment." Let's look at it now. To move as quickly as possible, I'll use a very simple example again. It is not true that "All ravens are black" implies that you will see any black ravens. All it does on its own is forbid observations of other kinds of ravens. If you believe that some particular bird is a raven with its natural coloration intact, and also that all ravens are black, then you can infer a prediction about how things should look. But even in this case, the prediction comes from more than the generalization itself.

Suppose a prediction made by a theory together with background assumptions does work out. Perhaps Newton's laws of motion and gravity, together with other assumptions, are used to successfully predict an observation about what we see in the night sky—the return of a comet, for example. Then it seems that the whole network of claims operating here, the laws and the other assumptions, all get some credit from the prediction turning out well. However, we can add some apparently irrelevant claims to Newton's laws and still get the same prediction. Newton's laws plus Freud's psychology, along with the background assumptions we needed before, will give us the same successful prediction about the sky. Do we give some credit to Freud as well? One might reply that Freud's psychology played no role here. It should get no credit. We do just as well, in this case, with Freud in the picture and without him. However, quite a lot of the content of Newton's physics also played no role in the prediction. Newton's laws cover all physical bodies, not just comets and stars and planets. Should we give credit only to some parts of Newton's theory, when it does well with the comet? A reply: "But Newton's laws don't distinguish comets from other physical bodies, and that is part of what made his theory so good." Right, but you don't need this aspect of his view to derive predictions about the comet.

Perhaps we might just accept that only the part of Newton's theory that we need to make the prediction should get any credit. But the bit of the theory we need might turn out to be rather small. We have ended up in a surprising place yet again. Despite all this, it is hard to believe that there is nothing right in the HD approach.

Further Reading and Notes

Hume's classic discussions of induction are in his *Treatise of Human Nature* ([1739] 1978) and his shorter and more readable *Enquiry Concerning Human Understanding* (1748). Once again, Hempel's *Aspects of Scientific Explanation* (1965) is a central source for logical empiricist ideas on this topic. Skyrms's *Choice and Chance* (2000) is a good introductory book on these issues, and it introduces probability theory as well. For Bacon, see Klein and Giglioni (2016).

For a discussion of explanatory inference, see the Harman paper I cited (1965), Peter Lipton's *Inference to the Best Explanation* (1991), and Achinstein's *The Book of Evidence* (2001). For "abduction," see Peirce ([1898] 1993). There are several different ways of using order-of-observation to address the ravens problem. The version that influenced me is in Horwich's *Probability and Evidence* (1982). The Wason/Hempel connection has been noted independently by a number of people. The first may have been Yael Cohen (1987).

Goodman's classic presentation of his "new riddle of induction" is in *Fact, Fiction and Forecast* (1955). The problem is in chapter 3 (along with other interesting ideas), and his solution is in chapter 4. Douglas Stalker has edited a collection on Goodman's riddle, called *Grue!* (1994). The Quine and Jackson chapters are particularly good.

Chapter 4

Popper

Conjecture and Refutation

4.1 *Popper's Unique Place in the Philosophy of Science*

Karl Popper is the only philosopher discussed in this book who is regarded as a hero by many scientists. Attitudes toward philosophy among scientists vary, but hardly ever does a philosopher succeed in inspiring scientists in the way Popper has. It is also fairly rare for a philosopher's view of science to be used within a scientific debate to justify one position over another. This too has happened with Popper. Within biology, debates about the classification of organisms and about ecology have seen Popper's ideas used in this way (Hull 1999). In 1965, Karl Popper became Sir Karl, knighted by the queen of England.

Popper's appeal is not surprising. His view of science is centered around a couple of simple, clear, and striking ideas. His vision of the scientific enterprise is a rather heroic one. Popper's theory of science has been criticized a great deal by philosophers over the years. I agree with many of these criticisms and don't see any way for Popper to escape their force. Despite the criticism, Popper's views continue to have an important place in philosophy—he saw some things that others seemed not to see—and his ideas continue to appeal to many working scientists.

4.2 *Popper's Theory of Science*

Popper began his intellectual career in Vienna, between the two world wars. He was not part of the Vienna Circle, but he did have contact with the logical positivists. Like the logical positivists, Popper left Europe upon the rise of Nazism, and after spending the war years in New Zealand, he moved to the London School of Economics, where he remained for the rest of his career. There he built a loyal group of allies, whom he often accused of disloyalty. His seminar series at the London School of Economics became famous for its grueling questioning, and for the fact that speakers had a difficult time presenting much of their lectures because of Popper's interruptions. Popper once had a famous confrontation with Wittgenstein, on the latter's turf at Cambridge University. One version of the story, told by Popper himself, has Wittgenstein brandishing a fireplace poker during a discussion of ethical rules, leading Popper to give an example of an ethical rule: "not to threaten visiting lecturers with pokers." Wittgenstein stormed out. Other versions of the story, including those told by Wittgenstein's allies, deny Popper's account.

The logical positivists developed their theory of science as part of a general theory of language, meaning, and knowledge. Popper was not much interested in these broader topics, at least initially; his primary aim was to understand science. As his first order of business, he wanted to understand the difference between scientific and nonscientific theories. In particular, he wanted to distinguish science from "pseudo-science." He did not regard pseudo-scientific ideas as meaningless; they just weren't science. For Popper, an inspiring example of genuine science was the work of Einstein. Examples of pseudo-science were Freudian psychology and Marxist views about society and history.

Popper called the problem of distinguishing science from non-science the "problem of demarcation." All of Popper's philosophy starts from his proposed solution to this problem. "Falsificationism" was the name Popper gave to his solution. Falsificationism claims that *a hypothesis is scientific if and only if it has the potential to be refuted by some possible observation*. To be scientific, a hypothesis has to take a risk, has to "stick its neck out." If a theory takes no risks at all, because it is compatible with every possible observation, then it is not scientific. As I said above, Popper held that Marx's and Freud's theories were not scientific in this sense. No matter what happens, Popper thought, a Marxist or Freudian can fit it into their theory somehow. These theories are never exposed to any risks.

So far I have described Popper's use of falsifiability to distinguish scientific from nonscientific theories. Popper also made use of the idea of falsification in a more far-reaching way. He claimed that all testing in science has the form of attempting to refute theories by means of observation. Crucially, for Popper it is never possible to confirm or establish a theory by showing its agreement with observations. Confirmation is a myth. The only thing an observational test can do is to show that a theory is false. So the truth of a scientific theory can never be supported by observational evidence, not even a little bit, and not even if the theory makes a huge number of predictions that all come out as expected.

As you might expect, Popper was a severe critic of the logical empiricists' attempts to develop a theory of confirmation or inductive logic. The problems they encountered, some of which I discussed in chapter 3, were music to his ears. Popper, like Hume, was an inductive skeptic, and Popper was skeptical (at least officially) about all forms of confirmation and support other than deductive logic itself.

Skepticism about induction and confirmation is a much more controversial position than Popper's use of falsification to solve the demarcation problem. Most philosophers of science have thought that if induction and confirmation are myths, that is very bad news for science. Popper tried to argue that there is no reason to worry; induction is a myth, but science does not need it anyway. Inductive skepticism, for Popper, is no threat to the rationality of science. In the opinion of most philosophers, Popper's attempt to defend this radical claim was not successful, and some of his discussions of this topic are rather misleading to readers. Some of the scientists who regard Popper as a hero do not realize that Popper believed it is never possible to confirm a theory, not even slightly, and no matter how many observations the theory helps us to predict successfully.

Popper placed great emphasis on the idea that we can never be completely sure that a theory is true. After all, Newton's physics was viewed as the best-supported theory in science, but early in the twentieth century it was shown to be false in several respects. However, almost all philosophers of science accept that we can never be 100 percent certain about factual matters, especially those discussed in science. This position, that we can never be completely certain about factual issues, is often known as *fallibilism* (a term due to C. S. Peirce). Most philosophers of science accept fallibilism. The harder question is whether or not we can be reasonable in increasing our confidence in the truth of a theory when it passes observational tests. Popper said no. The logical empiricists and most other philosophers of science say yes.

Popper had a fairly simple view of how testing in science proceeds. We take a theory that someone has proposed, and deduce an observational prediction from it. We then check to see if the prediction comes out as the theory says it will. If the prediction fails, then we have refuted, or falsified, the theory. If the prediction does come out as predicted, then all we should say is that we have not yet falsified the theory. For Popper, we cannot conclude that the theory is true, or that it is probably true, or even that it is more likely to be true than it was before the test. The theory *might* be true, but we can't say more than that.

We then try to falsify the theory in some other way, with a new prediction. We keep doing this until we have succeeded in falsifying it. What if years pass and we seem to never be able to falsify a theory, despite repeated tests? We can say that the theory has now survived repeated attempts to falsify it, but that's all. We never increase our confidence in the truth of the theory, and ideally, we never stop trying to falsify it. That's not to say we should spend all our time testing theories that have passed tests over and over again. We do not have the time and resources to test everything that could be tested. But that is just a practical constraint. According to Popper, we should always retain a tentative attitude toward our theories, no matter how successful they have been in the past.

In defending this view, Popper made much of the difference between confirming and disconfirming statements of scientific law. If someone proposes a law of the form "All *F*s are *G*," all it takes is one observation of an *F* that is not a *G* to falsify the hypothesis. This is a matter of deductive logic. But it is never possible to assemble enough observations to conclusively demonstrate the truth of such a hypothesis. You might wonder about situations where there are only a small number of *F*s and we could hope to check them all. Popper and the logical empiricists regarded these as unimportant situations that do not often arise in science. Their aim was to describe testing in situations where there is a huge or infinite number of cases covered by a hypothesized law or generalization. So Popper stressed that universal statements are hard or impossible to verify but easy, in principle, to falsify. The logical empiricist might reply that statements of the form "Some *F*s are *G*" have the opposite feature; they are easy to verify but hard or impossible to falsify. But

Popper claimed (and the logical empiricists tended to agree) that real scientific theories rarely take this form, even though some statements in science do.

Despite insisting that we can never support or confirm scientific theories, Popper believed that science is a search for true descriptions of the world. How can one search for truth if confirmation is impossible? This is an unusual kind of search. We might compare it to a certain kind of search for the Holy Grail, conducted by an imaginary medieval knight.

Suppose there are lots of grails around, but only one of them is holy. In fact, the number of non-holy grails is infinite or enormous, and you will never encounter them all in a lifetime. All the grails glow, but only the Holy Grail glows forever. The others eventually stop glowing, but there is no telling when any particular non-holy grail will stop glowing. All you can do is pick up one grail and carry it around and see if it keeps on glowing. You are only able to carry one at a time. If the one you are carrying is the Holy Grail, it will never stop glowing. But you would never know if you currently had the Holy Grail, because the grail you are carrying might stop glowing at any moment. All you can do is reject grails that are clearly not holy (since they stop glowing at some point) and keep picking up a new one. You will eventually die (with no afterlife, in this scenario) without knowing whether you succeeded.

This is similar to Popper's picture of science's search for truth. All we can do is try out one theory after another. A theory that we have failed to falsify up till now might be true. But if so, we will never know this or even have reason to increase our confidence.

Here is a quote from Popper's most important work, *The Logic of Scientific Discovery*, expressing his view:

> I think that we shall have to get accustomed to the idea that we must not look upon science as a "body of knowledge," but rather as a system of hypotheses; that is to say, as a system of guesses or anticipations which in principle cannot be justified, but with which we work as long as they stand up to tests, and of which we are never justified in saying that we know they are "true" or "more or less certain" or even "probable." (1959, 317)

4.3 *Popper on Scientific Change*

So far I have described Popper's views about the demarcation of science from non-science and the nature of scientific testing. Popper also used the idea of falsification to propose a theory of scientific change.

Popper's theory has an appealing simplicity. Science changes via a two-step cycle that repeats endlessly. Stage 1 in the cycle is *conjecture*—a scientist will offer a hypothesis that might describe and explain some part of the world. A good conjecture is a bold one, one that takes a lot of risks by making novel predictions. Stage 2 in the cycle is *attempted refutation*—the hypothesis is subjected to critical testing, in an attempt to show that it is false. Once the hypothesis is refuted, we go back to stage 1 again—a new conjecture is offered. That is followed by stage 2, and so on.

As the process moves along, it is natural for a scientist to propose conjectures that have some relation to previous ones. A theoretical idea can be refined and modified via many rounds of conjecture and refutation. That is fine, for Popper, though it is not essential. One thing that a scientist should not do, however, is react to the falsification of one conjecture by cooking up a new conjecture that is designed to just avoid the problems revealed by earlier testing, and that goes no further. We should not make ad hoc moves that merely patch the problems found in earlier conjectures. Instead, a scientist should constantly strive to increase the breadth of application of a theory, and increase the precision of its predictions. That means constantly trying to increase the "boldness" of conjectures.

What sort of theory is this? Popper intended it as a description of the general pattern that we actually see in science, and as a description of good scientific behavior as well. He accepted that not all scientists succeed in sticking to this pattern of behavior all the time. Sometimes people become too wedded to their hypotheses and refuse to give them up when testing tells them to. But Popper thought that a lot of actual scientific behavior does follow this pattern and that we see it especially in great scientists such as Einstein. For Popper, a good or great scientist is someone who combines two features, one corresponding to each stage of the cycle. The first feature is an ability to come up with imaginative, creative, and risky ideas. The second is a hard-headed willingness to subject these imaginative ideas to rigorous critical testing. A good scientist has a creative, almost artistic, streak and a tough-minded, no-nonsense streak. Imagine a hard-headed cowboy out on the range, with a Stradivarius violin in his saddlebags. (Perhaps at this point you can see some of the reasons for Popper's popularity among scientists.)

Popper's view on this issue can apparently be applied in the same way to individuals and to groups of scientists. An isolated individual can behave scientifically by engaging in the process of conjecture and refutation. And a collection of scientists can each, at an individual level, follow Popper's two-step procedure. But it seems that another possibility is a division of labor; one individual (or team) comes up with a conjecture, and another attempts to refute it. The basic idea of a conjecture-and-refutation pattern seems compatible with all these possibilities. But the case where individual A does the conjecture and individual B does the refutation will be suspicious to Popper. If individual A is a true scientist, they should take a critical attitude toward their own ideas. If individual A is completely fixated on their conjecture, and individual B is fixated on showing that A is wrong in order to advance a different conjecture, this is not good scientific behavior, according to Popper.

This raises an interesting question. Empiricist philosophies emphasize the virtues of open-mindedness, and Popper's view, which I see as an unorthodox version of empiricism, is no exception. But perhaps an open-minded community can be built out of a collection of rather closed-minded individuals. If actual scientists are wedded to their own conjectures, but each is wedded to a different conjecture and would like to prove the others wrong, shouldn't the overall process of conjecture and refutation work? What is wrong with the situation where B's role is to critically test A's ideas? So long as the testing occurs, what does it matter whether A or B does it? One possible problem here is that if everyone is closed-minded in this way, the results of tests might have no impact on what people believe. Real openness in the community would require that falsifications retain their bite once they have been achieved.

We can also imagine another kind of division of labor, with specialists on each side of the Popperian combination—specialist conjecturers and refuters. To some extent, we do surely see something like this. All this presses the following question on Popper: if a combination of conjecture and refutation is what is characteristic of science, then why isn't it enough to have some combination of socially organized behaviors that gives rise to openness in the community as a whole?

Although Popper did take an interest in community standards within science, he did seem to have a picture in which the good scientist should, as an individual, be willing to perform both the imaginative and the critical roles. Good scientists should retain a tentative attitude toward all theories, including their own: "whenever we propose a solution to a problem, we ought to try as hard as we can to overthrow our solution, rather than defend it" (1959, p. xix).

This relates to the inspirational role of Popper's ideas in science, which continues to this day. A few years ago I was driving down a highway and heard on the radio part of a long interview with a successful biologist looking back on his career, talking about how he came to do work that made a difference. He gave a lot of credit to Popper for getting him to think the right way about science. Here is what he emphasized most. Popper taught him that it is *OK to be wrong*. It is good to be wrong, if you and others can learn from the error. He came back to this idea several times in the interview; it really seemed to affect how he approached his work.

I agree that Popper was onto something here, and there seems to be something healthy in the Popperian attitude. "Seems," I said—so far this is just an impression. Soon in this book we'll encounter arguments that this attitude might not be as helpful for science as it appears.

I will make one more point before moving on to criticisms of Popper. The two-step process of conjecture and refutation that Popper describes has a striking resemblance to another two-step process: Darwin's explanation of biological evolution in terms of *variation* and *natural selection*. In science, according to Popper, scientists generate conjectures that are subjected to critical testing. In evolution, according to both Darwin himself and more recent versions of evolutionary theory, populations evolve via a process in which variations appear in organisms in a random or "undirected" way, and these novel characteristics are tested through their effects on the organism in its interactions with the environment. Variations that help organisms to survive and reproduce, and that are of the kind that gets passed on in reproduction, tend to be preserved and become more common in the population over time.

Ironically, at one time Popper thought that Darwinism was not a scientific theory, but he later retracted that claim. In any case, Popper and others have explored in detail the analogy between Popperian science and Darwinian evolution. The analogy should not be taken too seriously; evolution is not a process in which populations really "search" for anything, in the way that scientists search for good theories, and there are other important differences too. But the similarity is certainly interesting.

4.4 *Objections to Popper on Falsification*

I now turn to a critical assessment of Popper's ideas, beginning with his solution to the demarcation problem. Is falsifiability a good way to distinguish scientific ideas from nonscientific ones?

Let me first say that I think this question probably has no answer in the form in which Popper expressed it. We should not expect to be able to go through a list of statements or theories and label them "scientific" or "not scientific." However, I suggest that something fairly similar to Popper's question about demarcation does make sense: can we describe a distinctive scientific *strategy* of investigating the world, a scientific way of *handling* ideas?

Some of Popper's ideas are useful in trying to answer this question. In particular, Popper's claim that scientific theories should take *risks* is a good one. But Popper had an overly simple picture of how this risk-taking works. For Popper, theories have the form of generalizations, and they take risks by prohibiting certain kinds of particular events from being observed. If we believe that all pieces of iron, of whatever size and shape, expand when heated, then our theory forbids the observation of something that we know to be a piece of iron contracting when heated. A problem may have occurred to you: how sure can we be that, if we see a piece of "iron" contracting when heated, that it is really iron? We might also have doubts about our measurements of the contraction and the temperature change. Maybe the generalization about iron expanding when heated is true, but our assumptions about the testing situation and our ability to know that a sample is made of iron are false.

This problem is a reappearance of an issue discussed in chapter 2: holism about testing. Whenever we try to test a theory by comparing it with observations, we must make a large number of additional assumptions in order to bring the theory and the observations into contact with each other. If we want to test whether iron always expands when heated, we need to make assumptions about our ability to find or make reasonably pure samples of iron. If we want to test whether the amounts of the bases C and G are equal and the amounts of A and T are equal in all samples of DNA (Chargaff's rules, from chapter 3), we need to make a lot of assumptions about our chemical techniques. If we observe an unexpected result (iron contracting on heating, twice as much C as G in a sample of DNA), it is always possible to blame one of these extra assumptions rather than the theory we are trying to test. In extreme cases, we might even claim that the apparent observation was completely misunderstood or wrongly described by the observers. Indeed, this is not so uncommon in our attempts to work out what to make of reports of miracles and UFO abductions. So how can we really use observations to falsify theories in the way Popper wants? This is a problem not just for Popper's solution to the demarcation problem, but for his whole theory of science as well.

Some familiar ways of talking about testing—seen in Popper and also others—can be quite misleading on this front. People often say: a theory implies an observational prediction. Or, as in Popper's demarcation criterion, a hypothesis is only scientific if it has the potential to be refuted by some possible observation that clashes with it. It is never as simple as this, even in the ultra-simple "All ravens are black" cases. (I discussed this also in the optional section at the end of the previous chapter.) A generalization can tell you that if some object is a raven, then it has to be black. But working out whether there are any ravens around is a separate matter. The generalization plus some other assumptions tells you what you should see. And when the observation that comes back is a surprising one, there is always more than one possible thing to blame. A theory can't "take risks," in the way Popper likes, all on its own.

Popper was aware of this problem, and he struggled with it. He regarded the extra assumptions needed to connect theories with testing situations as scientific claims that might well be false—these are conjectures too. We can try to test these conjectures separately. But Popper conceded that logic itself can never force a scientist to give up a particular theory in the face of surprising observations. Logically, it is always possible to blame other assumptions involved in the test. Popper thought that a good scientist would not try to do this; a good scientist would want to expose the theory itself to tests and not try to deflect the blame.

Does this answer the holist objection? What Popper has done is move from describing a characteristic of scientific *theories* to describing a characteristic of scientific *behavior*. In some ways this is a retraction of his initial aim, which was to describe something about scientific theories themselves that makes them special. That is a problem. Then again, this shift to describing scientific modes of thought and behavior, rather than theories, might be a step forward. I'll have a closer look at this idea later in this chapter.

Popper also accepted that we cannot be completely certain about the observation reports that we use to falsify theories. We have to regard the acceptance of an observation report as a "decision," one that is freely made. Once we have made the decision, we can use the observation report to falsify any theory that conflicts with it. But for Popper, any falsification process is based, in the end, on a decision that could be challenged. Someone might come along later and try to show, via more testing, that the observation report was not a good one; this person might investigate whether the conditions of observation were misleading. That testing will have the same conjecture-and-refutation form described earlier. So the investigation into the controversial observation ultimately depends on "decisions" too.

Is this bad news for Popper? Popper insisted that making these decisions about single observations is very different from making free decisions directly about the theories themselves. But what sort of difference is this? If observation reports rest on nothing more than decisions, and these determine our choice of theories, how is that better than directly choosing the theories themselves, without worrying about observation? And couldn't we just "decide" to hang onto a theory and reject the observation reports that conflict with it? I am not saying that we should do these things, just that Popper has not given us a good reason not to do them. I believe that we should not do these things because we have good reason to believe that observation is a generally reliable way of forming beliefs, at least of some particular kinds. As I will argue in chapter 9, we can make use of a scientific theory of perception at this point in the story. But that argument will have to come later. Popper himself does not try to answer these questions by giving an argument about the reliability of perception.

This point about the role of decisions affects Popper's ideas about demarcation as well as his ideas about testing. Any system of hypotheses can be held onto despite apparent falsification if people are willing to make certain decisions. Given this, does Popper's view end up giving us any way to differentiate between science and pseudo-science? The answer is "yes and no." The "no" comes from the fact that scientific theories can be handled in a way that makes them immune to falsification, and nonscientific theories can be rejected if people decide to accept claims about particular matters that are incompatible with the theory. But there is a "yes" part in the answer as well. A scientific theory is falsifiable via a certain *kind* of decision—a decision about an observation report, which together with background assumptions can clash with a theory. A pseudo-scientific theory, Popper says, does not have this feature. So if a pseudo-scientific theory is to be rejected, some different kind of decision must be made.

There is another problem with Popper's views about falsification to discuss. The problem is bad for Popper, but I should emphasize that it is bad for many others as well. What can Popper say about theories that do not claim that some observation is forbidden, but only that it is very *unlikely*? If I believe that a certain coin is "fair," I can deduce from this hypothesis various claims about the probabilities of long all-heads or all-tails sequences of tosses. Suppose I observe a hundred tosses turning up heads a hundred times. This is very unlikely, according to my hypothesis about the coin, but it is not impossible. Any finite stretch of heads tosses is possible with a fair coin, although longer and longer runs of heads are treated by the theory as more and more unlikely. But if a hypothesis does not forbid any particular observations, then, according to Popper, it is taking no risks. This seems to entail that for Popper, theories that ascribe very low probabilities to specific observations, but do not rule them out altogether, are unfalsifiable and hence unscientific.

Popper's response was to accept that, logically speaking, hypotheses of this kind are unscientific. But this seems to make a mockery of the important role of probability in science. So Popper said that a scientist can decide that if a theory claims that a particular observation is extremely improbable, the theory *in practice* rules out that observation. If the observation is made, the theory is, in practice, falsified. According to Popper, it is up to scientists to work out, for their own fields, what sort of probability is so low that events of that kind are treated as prohibited. Probabilistic theories can only be construed as falsifiable in a special "in practice" sense. And we have here another role for decisions in Popper's philosophy of science, as opposed to the constraints of logic.

Popper is right that scientists reject theories when observations occur that the theory says are highly improbable (although which kinds of improbability have this importance is a complicated matter). And Popper is right that scientists spend a good deal of time working out "how improbable is *too* improbable." Complex statistical methods are used to help scientists with these decisions. But in making this move, Popper has damaged his original picture of science. This was a picture in which observations, once accepted, have the power to decisively refute theoretical hypotheses. That is a matter of deductive logic, as Popper endlessly stressed (though, as we saw, this has to work with the aid of background assumptions). Now Popper is saying that falsification can occur without its being backed up by a deductive logical relation between observation and theory.

4.5 *Objections to Popper on Confirmation*

As described earlier, Popper believed that theories can never be confirmed by observations, and he thought that inductive arguments are never justified. Popper thought that a theory of the rational choice of theories could be given entirely in terms of falsification, so he thought that rejecting induction and confirmation was no problem.

In the previous section I discussed problems with Popper's views about falsification. But let us leave those problems aside now, and assume in this section that we can use Popperian falsification as a method for decisively rejecting theories. If we make this assumption, is Popper's attempt to describe rational theory choice successful?

Here is simple problem that Popper has a very difficult time with. Suppose we are trying to build a bridge, and we need to use physical theories to tell us which designs are stable and will support the weight that the bridge must carry. This is a situation in which we must apply our scientific theories to a practical task. As a matter of fact, engineers and scientists in this situation will undoubtedly tend to use physical theories that have survived empirical testing; they will use "tried and true" methods as far as possible. The empiricist approach to the philosophy of science holds that such a policy is rational. A problem for empiricism is to explain in more detail why this policy is the right one. That task is hard, as I hope became clear in chapter 3. But let us focus on Popper, who wants to avoid the need for a theory of confirmation. How does Popper's philosophy treat the bridge-building situation?

Popper can say something about why we should prefer to use a theory that has not been falsified over a theory that has been falsified. Theories that have been falsified have been shown to be false (here again I ignore the problems discussed in the previous section). But suppose we have to choose between (1) a theory that has been tested many times and has passed every test and (2) a brand-new theory that has just been conjectured and has never been tested. Neither theory has been falsified. We would ordinarily think that the rational thing to do is to choose the theory that has survived testing. But what can Popper say about this choice? Why exactly would it be irrational, for Popper, to build the bridge using a new theory that has not yet been tested?

Popper recognized and struggled with this problem too. Perhaps this has been the most common objection to Popper from other empiricist philosophers (e.g., Salmon 1981). Popper is unable to give a very good reply. Popper refuses to say that when a theory passes tests, we have more reason to believe that the theory is true. Both the untested theory and the well-tested theory are just conjectures. But Popper did devise a special concept to use in this situation. He said that a theory that has survived many attempts to falsify it is "corroborated." And when we face choices like the bridge-building one, it is rational to choose corroborated theories over theories that are not corroborated.

What is corroboration? Popper gave a technical definition and held that we can measure the amount of corroboration that a theory has at a particular time. The technicalities do not matter, though. We need to ask, what *sort* of property is corroboration? Has Popper just given a new name to confirmation? If so, he can answer the question about building the bridge, but he has given up one of the main features that distinguishes his view from the logical empiricists and everyone else. If corroboration is totally different from confirmation—so different that we cannot regard corroboration as any guide to a theory's truth—then why should we choose a corroborated theory when we build the bridge?

This issue has been much discussed (see Newton-Smith 1981). Popper's concept of corroboration can be interpreted in a way that makes it different from confirmation, but then Popper can give no good answer to the question of why we should choose corroborated theories over new ones when building bridges.

To understand corroboration, think of the difference between an academic transcript and a letter of recommendation. This distinction should be vivid to students! An academic transcript says what you have done. It measures your past performance, but it does not contain explicit predictions about what you will do in the future. A letter of recommendation usually says something about what you have done, and it also makes claims about how you are likely to do in the future. Confirmation, as understood by the logical empiricists, is something like a letter of recommendation for a scientific theory. Corroboration, for Popper, is like an academic transcript. And Popper thought that no good reasons could be given for believing that past performance is a reliable guide to the future. So corroboration is entirely backward-looking. Consequently, no reason can be given for building a bridge with a corroborated theory rather than a noncorroborated but unfalsified one.

I think the best thing for Popper to say about the bridge-building situation is to stick to his inductive skepticism. He should argue that we really don't know what will happen if we build another bridge with a design that has worked in the past. Maybe it will stay up and maybe it won't. There might also be practical reasons for choosing that design if we are very familiar with it. But if someone comes along with a brand-new, untested design, we won't know whether it's a bad design until we try it. Popper liked to say that there is no alternative policy that is *more* rational than using the familiar and well-tested design, and we do have to make some decision. So we can go ahead and use the established design. But as Wesley Salmon (1981) replied, this does not help at all. If confirmation does not exist, then it seems there is also no policy that is more rational than choosing the *un*tested design. All we have here is a kind of "tie" between the options.

For most people, this is an unsatisfactory place for a philosophy of science to end up. Inductive skepticism of this kind is hard to take seriously outside of abstract, academic discussion. However, the efforts seen over several centuries have shown how hard it is to produce a good theory of induction and confirmation. One of the valuable roles of Popper's philosophy is to show what sort of theory of science might be possible if we give up on induction and confirmation.

In the first chapter of this book, I said that few philosophers still try to give descriptions of a definite "scientific method," where this is construed as something like a recipe for science. Popper is a partial exception here, since he does come close to giving a kind of recipe (although Popper insists there is no recipe for coming up with interesting conjectures). His view has an interesting relationship to descriptions of scientific method given in science textbooks.

In many textbooks, one finds something called the "hypothetico-deductive method." Back in chapter 3, I discussed a view about confirmation that is often called "hypothetico-deductivism." Now we are dealing with a method rather than a theory of confirmation. Science textbooks are more cautious about laying out recipes for science than they used to be, but descriptions of the hypothetico-deductive method are still fairly common. Formulations of the method vary, but some are basically a combination of Popper's view of testing and a less skeptical view about confirmation. In these versions, the hypothetico-deductive method is a process in which scientists come up with conjectures and then deduce observational predictions from those conjectures. If the predictions come out as the theory says, then the theory is supported. If the predictions do not come out as the theory says, the theory is not supported and should be rejected. This process has the basic pattern that Popper described, but the idea that theories can be "supported" by observations is not a Popperian idea.

The term "support" is vague, but I think discussions of the hypothetico-deductive method generally assume that if a theory makes a lot of successful predictions, we have more reason to believe that the theory is true than we had before the successful predictions were made. We will never be completely sure, but the more tests a theory passes, the more confidence we can have in its truth. The idea that we can gradually increase our confidence that a theory is true is an idea that Popper rejected. As I said at the start of this chapter, some of Popper's scientific admirers do not realize that Popper's view has this feature, because some of Popper's discussions were misleading.

Other formulations of the hypothetico-deductive method include a first stage in which observations are collected and a conjecture is generated from these observations. Popper disagreed with this picture of scientific procedure because he argued that fact-gathering always takes place in a way guided by conjectures. But this is a fairly minor point.

Another term that some textbooks use in discussing scientific method is "strong inference." This term was introduced by a chemist named John Platt (1964). Strong inference is roughly a Popperian kind of testing together with a further assumption, which Popper rejected. This assumption is that we can write down all the possible theories that might be true in some area, and test them one by one. We find the true theory by eliminating the alternatives. For Popper, this is impossible; in any real case, there will be an infinite number of competing theories. So even if we eliminate ten or a hundred possibilities, the same infinite number still remains. According to Popper, all we can do is to choose one theory, test it, then choose another, and so on. We can never have confidence that we have eliminated all, or most, of the alternatives.

I have not discussed objections to Popper's theory of scientific change yet, but I will do so in the next few chapters.

What is Popper's single most important and enduring contribution to philosophy of science? I'd say it is his use of the idea of *riskiness* to describe the kind of contact that scientific theories have with observation. Popper was right to concentrate on the ideas of exposure and risk in his description of science. Science tries to formulate and handle ideas in such a way that they are exposed to falsification and modification via observation. Popper's formulation is valuable because it captures the idea that theories can appear to have lots of contact with observation when in fact they only have a kind of "pseudo-contact" with observation, because they avoid all risks. This idea is a real advance. Popper's analysis of how this exposure works has a lot of problems, but the basic idea is good.

4.6 *Further Comments on the Demarcation Problem*

Popper is on to something when he says that scientific theories should take risks. In this section I will try to develop this idea a bit differently. Popper was interested in distinguishing scientific theories from unscientific ones, and he wanted to use the idea of risk-taking to make the distinction. But this idea of risk-taking is better used as a way of distinguishing scientific from unscientific ways of handling ideas. And we should not expect a sharp distinction between the two.

The scientific way of handling an idea is to try to connect it with other ideas, to embed it in a larger conceptual structure, in a way that *exposes it to observation*. This "exposure" is not a matter of simple falsification; there are many ways in which exposure to observation can be used to modify and assess an idea. But if a hypothesis is handled in a way that keeps it apart from all the risks associated with observation, that is an unscientific handling of the idea.

So it is a mistake to try to work out whether theories such as Marxism or Freudianism are themselves "scientific" or not, as Popper did. A big idea like Marxism or Freudianism will have scientific and unscientific *versions*, because the main principles of the theory can be handled scientifically or unscientifically. Scientific versions of Marxism and Freudianism are produced when the main principles are connected with other ideas in a way that exposes these principles to testing. To scientifically handle the basic principles of Marxism is to try to work out what difference it would make to things we can observe if Marxist principles were true. To do this it is not necessary that we write down some single observation that, if we encounter it, will lead us to definitively reject the main principles of the theory. It will remain possible that a background assumption is at fault, and there is no simple recipe for adjudicating such decisions.

To continue with Popper's examples, Marxism holds that the driving force of human history is struggle between economic classes, guided by ongoing changes in economic organization. This struggle results in a predictable sequence of political changes, leading eventually to socialism. Freudianism holds that the normal development of a child includes a series of interactions and conflicts between unconscious aspects of the child's mind, where these processes include resolving sexual feelings toward their parents. Adventurous ideas like these can be handled scientifically or unscientifically. Over the last century, the Marxist view of history has been handled scientifically enough for it to have been disconfirmed. Too much has happened that seems to have little to do with class struggle; the ever-increasing political role of religious and cultural solidarity is an example (Huntington 1996). Capitalist societies have also adapted to economic tensions in ways that Marxist views about politics and economics do not predict. It remains possible to hang on to the main principles of Marxism, but fewer and fewer people handle the theory in that way anymore. Many still think that Marxism contains useful insights about economic matters, but the fundamental claims of the theory have not held up well.

Freudianism is another matter; the ideas are still popular in some circles, but not because of success under empirical testing. Instead, the theory seems to hang around because of its striking and intriguing character, and because of a subculture in fields such as psychotherapy and literary theory that guards the main ideas and preserves them despite their empirical problems. The theory is handled very unscientifically by those groups. Freud's theory is not taken seriously by most scientifically oriented psychology departments in research universities, but it is taking a while for this fact to filter out to other disciplines.

Sometimes people say that Freud's ideas were indeed scientifically successful, because the idea that our minds contain unconscious processes is alive and well in scientific psychology. In that very broad sense, yes, a part of Freud's view is still with us, and recent work in neuroscience has investigated the idea of unconscious thought processes in ways that are undeniably scientific (Dehaene 2014). But I don't think Freud should get much credit for that broader idea, and there is not much left in psychology of the little internal agents ("id," "superego," etc.) that Freud wanted to describe.

Evolution is another big idea that can be handled either scientifically or unscientifically. People (including Popper) have wondered from time to time whether evolutionary theory, or some specific version of it such as Darwinism, is testable. What observations would lead scientists to give up current versions of evolutionary theory? A one-line reply that biologists sometimes give to this question is "a Precambrian rabbit." J. B. S. Haldane, an important biologist of the early twentieth century, is often credited with the line. An evolutionary biology textbook by Douglas Futuyma expresses the same point more soberly: finding "incontrovertibly mammalian fossils in incontrovertibly pre-cambrian rocks" would "refute or cast serious doubt on evolution" (1998, 760). The one-liner is a start, but the real situation is more complicated. So let us look at the case.

The Precambrian ended around 540 million years ago (the term "Precambrian" covers a number of different periods in the history of the Earth, all before the Cambrian, which began about 540 million years ago). Suppose we found a well-preserved rabbit fossil in rocks 600 million years old. All our other evidence suggests that the only animals around then were soft-bodied invertebrates (many of them very strange indeed) and that mammals did not appear until over 300 million years later. Of course, a good deal of suspicion would be directed toward the finding itself. How sure are we that the rocks are that old? Might the rabbit fossil have been planted as a hoax? Remember the apparent fossil link between humans and apes that turned out to be a hoax, the "Piltdown Man" of 1908 (Feder 1996). Here we encounter another aspect of the problem of holism about testing—the challenging of observation reports, especially observation reports that are expressed in a way that presupposes other pieces of theoretical knowledge. But let us suppose that all agree the fossil is clearly a Precambrian rabbit.

This finding would not be an instant falsification of all of evolutionary theory, because evolutionary theory is now a diverse package of ideas, including abstract theoretical models as well as claims about the actual history of life on Earth. The theoretical models are intended to describe what various evolutionary mechanisms can do in principle. Claims of that kind are usually tested with mathematical analysis and computer simulation. Evolution on a small scale can also be observed directly in the lab, especially in bacteria, viruses, yeast, and some animals like fruit flies. The Precambrian rabbit would not affect those results. But a Precambrian rabbit fossil would show that somewhere in the package of central claims found in evolutionary biology textbooks, there are some serious errors. These would at least include errors about the overall history of life, about the kinds of processes through which a rabbit-like organism could evolve, and probably about the "family tree" of species on Earth. The challenge would be to work out where the errors lie, and that would require separating out and independently reassessing each of the ideas that make up the package. This reassessment could, in principle, result in the discarding of basic evolutionary beliefs, like the idea that humans evolved from nonhuman animals.

Over the past forty years or so, evolutionary theory has in fact been exposed to a huge and sustained empirical test, because of advances in molecular biology. Since the time of Darwin, biologists have been trying to work out the total "tree of life" linking all species on Earth, by comparing their similarities and differences and taking into account factors such as geographical distribution. The trees that were arrived at prior to the rise of molecular biology can be seen summarized in various picturesque old charts and posters.

More recently, molecular biology has made it possible to compare the DNA sequences of many species. Similarity in DNA is a good indicator of the closeness of evolutionary relationship. Claims about the evolutionary relationships between different species can be tested fairly directly by discovering how similar their DNA is and estimating how many years of independent evolution the different species have undergone since they last shared a common ancestor. As this work began, it was reasonable to wonder whether the wealth of new information about DNA would be compatible or incompatible with the family tree that had been worked out previously. Suppose the DNA differences between humans and chimps suggested that the human lineage split off from the lineage that led to chimps many hundreds of millions of years ago and that humans are instead closely genetically related to squid. This would have been an enormous shock for evolutionary theory, one of almost the same magnitude as the Precambrian rabbit.

As it happened, the DNA data suggest that humans and chimps diverged about six million years ago and that chimps, along with bonobos, are our nearest living relatives. Prior to the DNA data, it was unclear whether humans were more closely related to chimps or to gorillas, and the date for the chimp-human divergence was much less clear. That is how the grand test of our old pre-molecular family tree has tended to go. There have been lots of new discoveries, and a number of interesting adjustments to the old picture, but no huge surprises. The version of evolutionary theory that was worked out in the years before molecular genetics stood up pretty well.

Further Reading and Notes

Popper's most famous work is his book *The Logic of Scientific Discovery*, published in German in 1935 and in English in 1959. The book is mostly very readable. Chapters 1–5 and 10 are the key ones. For the issues in section 4.4 above, see chapter 5; for section 4.5, see chapter 10. A quick and clear introduction to Popper's ideas is the paper "Science: Conjectures and Refutations" in his collection *Conjectures and Refutations* (1963).

The *Cambridge Companion to Popper* (2016) is a collection of chapters reappraising Popper's ideas in different fields. It includes a paper of mine, "Popper's Philosophy of Science: Looking Ahead," that discusses some additional features of his view in a fairly positive light.

Newton-Smith's *The Rationality of Science* (1981) is an older book that is still useful for its clarity and its presentation of issues surrounding "corroboration." Salmon (1981) is an exceptionally good discussion of Popper's views on induction and prediction. Schilpp (1974) collects many critical essays on Popper, with Sir Karl's replies. For the story of Popper, Wittgenstein, and the brandished poker, see Edmonds and Eidinow (2001). Winther (2009) is a good discussion of prediction in evolutionary biology.

Popper's influence on biologists and his (often peculiar) ideas about evolution are discussed in Hull (1999). Horgan's book *The End of Science* (1996) contains a very entertaining interview with Popper. For a more informal exploration of the upsides and downsides of error in our beliefs, see Kathryn Schulz's *Being Wrong* (2010).

Chapter 5

Kuhn's Revolution

5.1 *"The Paradigm Has Shifted"*

In this chapter we encounter the most famous book about science written during the last hundred years: *The Structure of Scientific Revolutions*, by Thomas Kuhn. Kuhn's book was first published in 1962, and its impact was enormous. Just about everything written about science by philosophers, historians, and sociologists since then has been influenced by it. The book has also been hotly debated by scientists themselves. But *Structure* (as the book is known) has not only influenced these academic disciplines; many of Kuhn's ideas and terms have made their way into areas like politics and art as well.

A common way of describing the importance of Kuhn's book is to say that he shattered traditional myths about science, especially empiricist myths. Kuhn showed, according to this view, that actual scientific behavior has little to do with traditional philosophical theories of rationality and knowledge. There is some truth in this interpretation, but it is often exaggerated. Kuhn spent a good deal of his time after *Structure* trying to distance himself from some of the radical views of science that came after him, even though he was revered by the radicals. It may also come as a surprise to learn that Kuhn's book was published in a series organized and edited by the logical empiricists; *Structure* was published as part of their "International Encyclopedia of Unified Science" series. But there is no denying that this was something of a "Trojan horse" situation. Logical empiricism was widely perceived to have been greatly damaged by Kuhn.

I said above that some of Kuhn's ideas and terms have made their way into areas far from the philosophy of science. The best example is Kuhn's use of the term "paradigm." Here is a passage from Tom Wolfe's novel *A Man in Full*. Charlie Croker, a real-estate developer with debt problems, is talking with his financial adviser, Wismer ("the Wiz") Stroock:

> "I'm afraid that's a sunk cost, Charlie," said Wismer Stroock. "At this point the whole paradigm has shifted."
> Charlie started to remonstrate. Most of the Wiz's lingo he could put up with, even a "sunk cost." But this word "paradigm" absolutely

drove him up the wall, so much so that he had complained to the Wiz about it. The damned word meant nothing at all, near as he could make out, and yet it was always "shifting," whatever it was. In fact, that was the only thing the "paradigm" ever seemed to do. It only shifted. But he didn't have enough energy for another discussion with Wismer Stroock about technogeekspeak. So all he said was:

"OK, the paradigm has shifted. Which means what?" (1998, 71)

This sort of talk about "paradigm shifts" derives from Kuhn's book. But what is a paradigm? The short answer is that a paradigm, in Kuhn's theory, is a whole way of doing science, in some particular field. It is a package of claims about the world, methods for gathering and analyzing data, and habits of scientific thought and action. In Kuhn's theory of science, the big changes in how scientists see the world—the "revolutions" that science undergoes every now and then—occur when one paradigm replaces another. Kuhn argued that observational data and logic alone cannot force scientists to move from one paradigm to another, because different paradigms often include within them different rules for treating data and assessing theories. Some people have interpreted Kuhn as claiming that changes between paradigms are completely irrational, but Kuhn did not believe that. Instead, Kuhn had a rather complicated and subtle view about the roles of observation and logic in scientific change.

In a passage like the one from Wolfe above, "paradigm" is used in a looser way that is derived from its role in Kuhn's theory of science. A paradigm in this sense is something like a way of seeing the world and interacting with it.

Kuhn did not invent the word "paradigm." It was an established term that meant (roughly) an illustrative example on which other cases can be modeled. (Kuhn discusses this original meaning in *Structure*, on page 23). And although Kuhn's theory is the inspiration for all the talk about paradigm shifts that one hears, Kuhn only occasionally used the phrase "paradigm shift." More often he talked about paradigms changing or being replaced. Whichever term one uses, though, Kuhn's theory was itself something like a paradigm change in the history and philosophy of science. Nothing has been the same since.

5.2 *Paradigms: A Closer Look*

A moment ago I said that a paradigm, in Kuhn's theory, is a package of claims about the world, methods for gathering and analyzing data, and habits of scientific thought and action. However, it is more accurate to say that this is just one sense in which Kuhn used the term. In *Structure*, "paradigm" is used in a number of different ways. I will recognize two different senses of the term. The first sense, which I will call the *broad* sense, is the one I described above. Here, a paradigm is a package of ideas and methods, which, when combined, make up both a view of how some part of the world works, and a way of doing science. When I say "paradigm" in this book without adding "broad" or "narrow," I mean this broad sense. According to Kuhn, one part of a paradigm in the broad sense is a specific *achievement*, or an *exemplar*. This achievement might be a strikingly successful experiment, such as Mendel's experiments with peas, which eventually became the basis of modern genetics. It might be the formulation of a set of equations or laws, such as Newton's laws of motion or Maxwell's equations describing electromagnetism. Whatever it is, this achievement is a source of inspiration to others; it suggests a way to investigate the world. Kuhn often used the term "paradigm" for a specific achievement of this kind. I will call these paradigms in the *narrow* sense. Paradigms in the broad sense (whole ways of doing science) include within them paradigms in the narrow sense (examples that serve as models, inspiring and directing further work). Kuhn himself did not use this "narrow/broad" terminology, but it is helpful. When Kuhn first introduced the term "paradigm" in *Structure*, he defined it in the narrower sense. But in much of his writing, the broad sense is intended.

Kuhn used the phrase "normal science" for scientific work that occurs within the framework provided by a paradigm. A central feature of normal science is that it is well organized. Scientists doing normal science tend to agree on which problems are important, on how to approach these problems, and on how to assess possible solutions. They also agree on what the world is like, at least in broad outlines. A scientific revolution occurs when one paradigm breaks down and is replaced by another.

This initial sketch is enough for us to go straight to some central points about the message of Kuhn's book.

The first point can be approached via a contrast with Popper. For Popper, science is characterized by *permanent openness*, a permanent and all-encompassing critical stance, even with respect to the fundamental ideas in a field. Kuhn argued that it is false that science exhibits a permanent openness to the testing of fundamental ideas. Not only that, but science would be worse off if it had the kind of openness that philosophers have treasured.

The second point concerns scientific change. Here again a contrast with Popper is convenient. For Popper, all science proceeds via a single process, the process of conjecture and refutation. There can still be episodes called "revolutions" in such a view, but revolutions are just different in degree from what goes on the rest of the time; they involve bigger conjectures and more dramatic refutations. For Kuhn, there are two distinct kinds of scientific change: change within normal science, and revolutionary science. (These are bridged by "crisis science," a period of unstable stasis.) These two kinds of change have very different features; when we try to apply concepts such as *justification, rationality,* and *progress* to science, according to Kuhn we find that normal and revolutionary science have to be described differently. Within normal science, there are clear and agreed-upon standards for the justification of arguments; within revolutionary science there are not. Within normal science there is clear progress; within revolutionary science it is very hard to tell. Because revolutions are essential to science, the task of describing rationality and progress in science as a whole becomes complicated.

Before we go deeper into the details of Kuhn's view, there is one other preliminary point to make. This has to do with a question that one should always ask when thinking about Kuhn's theory and other theories like it. The question is, Which parts of the theory are just *descriptive*, and which are *normative*? That is, when is Kuhn just making a claim about how things are, and when is he making a value judgment, saying how they should be? Kuhn certainly accepted that he was making some normative claims (1996, 8). But it's often hard to tell when he is just saying how things are and when he is making claims about what is good or bad. My own interpretation of Kuhn stresses the normative element in his work. I think Kuhn had a definite picture of how science should work and of what can cause harm to it. In fact, it is here that we find what I regard as the most fascinating feature of *The Structure of Scientific Revolutions*. This is the relationship between

1. Kuhn's constant emphasis on the arbitrary, personal nature of factors often influencing scientific decisions, the rigidity of scientific indoctrination of students, the tenacity with which ideas are held by scientists and the "conceptual boxes" that nature gets forced into . . . , and
2. Kuhn's suggestion that these features are actually the key to science's *success*—without them, there is no way for scientific research to proceed as effectively as it does.

How can features that look like failings and flaws help science? How can it help science for decisions to be made on the basis of anything other than what the data say? To answer these questions, we need to look more closely at the details of Kuhn's story about scientific change.

5.3 *Normal Science*

Normal science is research inspired by a striking achievement that provides a basis for further work (a paradigm in the narrow sense). Kuhn does not think that all science needs a paradigm. Each scientific field starts out in a state of "pre-paradigm science." During this pre-paradigm state, scientific work can go on, but it is not well organized and usually not very effective.

At some point, however, some striking piece of work appears. This achievement is taken to provide insight into the workings of some part of the world, and it supplies a model for further investigation. This achievement is so impressive that a tradition of further work starts to grow up around it. The field has its first paradigm.

What are some examples of paradigms? Kuhn gave examples from physics and chemistry, such as Newton's and Einstein's paradigms. Here I will mention two cases from other fields. Within psychology around the middle of the twentieth century, a great deal of work was based upon the behaviorist approach of B. F. Skinner. Two basic principles of Skinnerian behaviorism are (1) that learning is basically the same in humans, rats, pigeons, and other animals and (2) that learning proceeds by reinforcement—behaviors followed by good consequences tend to be repeated, while behaviors followed by bad consequences tend not to be repeated (1938). Along with these principles, the Skinnerian paradigm included a set of experimental tools, such as an apparatus in which pigeons made choices in response to stimuli by pecking lighted keys. It also included statistical techniques used to analyze data and various habits and skills for working out relevant and interesting experiments.

Here is an example from biology. Modern molecular genetics is based on a set of principles such as: (1) genes are made of DNA (in all organisms except some viruses, which have RNA genes), (2) genes have their effects mostly by producing protein molecules, and (3) nucleic acids (DNA and RNA) specify the structure of proteins by determining the order of the units that make them up, and not vice versa. This last principle is often called "the central dogma" (Crick 1958). Along with these theoretical claims, molecular genetics includes a set of techniques for sequencing genes, for producing and studying mutations, for analyzing the similarity of different genes, and so on.

For Kuhn, a scientific field usually has only one paradigm guiding it at any particular time. Kuhn does allow that occasionally a field can be governed by several related paradigms, but this is rare. In general, Kuhn's picture has it that there is *one paradigm per field per time*.

A paradigm's role is to organize scientific work; the paradigm coordinates the work of individuals into an efficient collective enterprise. For Kuhn, a key feature that distinguishes normal science from other kinds is the absence of debate about fundamentals. Because scientists doing normal science agree on these fundamentals, they do not waste their time arguing about the most basic issues. Once biologists agree that genes are made of DNA, they can coordinate their work on how specific genes affect the characteristics of plants and animals. Once chemists agree that understanding chemical bonding is understanding the interactions between the outer layers of electrons within different atoms, they can work together to investigate when and how particular reactions will occur.

Kuhn places great emphasis on the "consensus-forging" role of paradigms. He argues that without it, there is no chance for scientists to achieve a deep understanding of phenomena. Detailed work and revealing discoveries require cooperation and consensus. Cooperation and consensus require closing off debate about fundamentals.

As usual, we should be careful to distinguish between the descriptive and the normative here. Kuhn certainly claims that normal science does close off debate about fundamentals. But does he go beyond that and claim this is something that normal science *should* do? I think Kuhn does think this (see Kuhn 1996, 24–25, 65), but the issues are controversial. If Kuhn does make a normative claim here, then we see an important contrast with Popper. Although Popper can certainly allow that not everything can be criticized at once, Popper's view does hold that a good scientist is permanently open-minded with respect to all issues in the field in which they are working, even the very basic issues. Popper criticized Kuhn explicitly on this point (1970); he said that although "normal science" of Kuhn's kind does sometimes occur, it is a bad thing that it does.

What is the work of a good normal scientist like? Kuhn describes much of the work done in normal science as "puzzle-solving." The normal scientist tries to use the tools and concepts provided by the paradigm to describe, model, or create new phenomena. The puzzle is trying to get a new case to fit smoothly into the framework provided by the paradigm. Kuhn used the term "puzzle" rather than "problem" for a reason. A puzzle is something we have not yet solved, but that we think does have a solution. A problem might, for all we know, have no solution. Normal science tries to apply the concepts provided by a paradigm to issues that the paradigm suggests should be soluble. Part of the guidance provided by a paradigm is guiding the selection of good puzzles.

The term "puzzle" also seems to suggest that the work is in some way insignificant or trivial. Here again, Kuhn intends to convey a precise message with the term. A normal scientist does, Kuhn thinks, spend a lot of time on topics that look insignificant from the outside. (He even uses the term "minuscule"; 1996, 24.) But it is this close attention to detail—which only the well-organized machine of normal science makes possible—that is able to reveal deep new facts about the world. I think Kuhn felt a kind of awe at the ability of normal science to home in on topics and phenomena that look insignificant from outside but that turn out eventually to have huge importance. And although the normal scientist is not trying to find phenomena that lead to paradigm change, these detailed discoveries often contain the seeds of large-scale change and the destruction of the paradigm that produced them.

5.4 *Anomaly and Crisis*

I said that a central feature of normal science, for Kuhn, is that the fundamental ideas associated with a paradigm are not debated. Normal scientists spend their time trying to extend the paradigm, theoretically and experimentally, to deal with new cases. When there is a failure to get the expected results, the good normal scientist reacts by trying to work out what mistake she or he has made. The proverb "only a poor workman blames his tools" applies.

Kuhn accepts that theories are sometimes refuted by observation; within normal science, hypotheses are refuted (and confirmed) all the time. The paradigm supplies principles for making these decisions. But throwing out an entire paradigm is much more difficult. According to Kuhn, the rejection of a paradigm happens only when (1) a critical mass of anomalies has arisen and (2) a rival paradigm has appeared. For now, we will look just at the first of these—the accumulation of a critical mass of anomalies.

An "anomaly" for Kuhn is a puzzle that has resisted solution. Kuhn holds that all paradigms face anomalies at any given time. As long as there are not too many of them, normal science proceeds as usual, and scientists regard them as a challenge. But the anomalies tend to accumulate. Sometimes a single one becomes particularly prominent by resisting the efforts of the best workers in the field. Eventually, according to Kuhn, the scientists start to lose faith in their paradigm. The result is a *crisis*.

Crisis science, for Kuhn, is a special period when an existing paradigm has lost the ability to inspire and guide scientists, but when no new paradigm has emerged to get the field back on track. For whatever reason, the scientists in a field lose their confidence in the paradigm. As a consequence, the most fundamental issues are back on the table for debate. Amusingly, Kuhn suggests that during crises scientists tend to suddenly become interested in philosophy, a field that he sees as quite useless for normal science.

I used the phrase "critical mass of anomalies" to describe the trigger for a crisis. This atomic-age metaphor is appropriate in several ways. In particular, I use it here to suggest that Kuhn sees the breakdown of a paradigm as something that is part of the "proper functioning" of science, though it does not feel that way to the scientists involved. Normal science is structured in a way that makes its own destruction inevitable, but only in response to the right stimulus. That stimulus is the appearance of problems that are deep rather than superficial, problems that reveal a real inadequacy in the paradigm. Because normal scientists will tolerate a good deal of temporary trouble without abandoning normal science, a paradigm does not break down easily. But when the right stimulus comes, the paradigm will disintegrate. In this way, a paradigm is like a well-shielded and well-designed bomb. A bomb is supposed to blow up; that is its function. But a bomb is not supposed to blow up at any old time; it's supposed to blow up in very specific circumstances. A well-designed bomb will be shielded from minor buffets. Only a very specific stimulus will trigger the explosion.

Some might find this militaristic analogy unpleasant, but I think it captures an important theme in Kuhn's work. All paradigms constantly encounter anomalies. For a Popperian, and many forms of empiricism, these anomalies should count as "refutations" of the theory. Kuhn thinks that science does not treat these ubiquitous anomalies as refutations, and it also should not. If scientists dropped their paradigms every time a problem arose, they would never get anything done.

Much of the secret of science, for Kuhn, is the balance it manages to strike between being too resistant to change and not being resistant enough. If the simplest form of empiricist thinking prevailed, people would throw ideas away too quickly when unexpected observations appeared, and chaos would result. Ideas need some protection, or they can never be properly developed. But if science were completely unresponsive to empirical failures, conceptual advance would grind to a halt. For Kuhn, science seems to achieve a delicate balance. This balance is not something we can describe in terms of a set of explicit rules. It exists implicitly in the social structures and transmitted traditions of scientific behavior, and in the quirks of the scientific mind.

This kind of balance, if it is real, involves interesting relationships between the properties of individuals and communities. We had a quick first look at this theme in the previous chapter. Popper wants to see open-mindedness, and an ongoing process of conjecture and refutation. I asked: might an open-minded *community* be built out of rather closed-minded *individuals*? If scientists are wedded to their own conjectures until refutations arrive, but each is wedded to a different conjecture and would like to prove the others wrong, shouldn't the process of conjecture and refutation work? What is wrong with the situation where *B*'s role is to critically test *A*'s ideas, without *A* being critical about their own ideas? In Kuhn we see a different sort of combination. Normal science is full of rather closed-minded individuals, usually with no one trying to knock over a paradigm. But by their intensely focused work and the exploration of anomalies, they produce the paradigm's collapse.

From this point in the book onward, these relations between features of individuals and communities will be an important theme. To help with this, I will introduce a three-way distinction between different perspectives on science. We can think of this as three scales or levels of analysis, from fine-grained to coarse-grained (see figure 5.1). There are not sharp boundaries between the levels, though, and it is also helpful to think of the situation using an analogy with a zoom lens on a camera —we can zoom in and out continuously.

First, there is a very fine-grained or zoomed-in perspective on science, where we are looking at the activities of individual people. Observation, reasoning, and belief are treated as features of the individual scientist. I will call this level 1.

If we zoom out a bit, we will find communities of scientists and their social networks. This is level 2. Here we see relationships where scientists

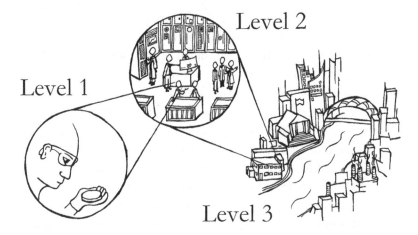

Figure 5.1. Three levels of analysis that can be applied to science, from the level of the individual (level 1), through the level of the scientific community (level 2), to the level of the whole society in which science is embedded (level 3)

collaborate and compete with each other. They use one another's work, criticize that work, and train newcomers.

As the camera zooms out further, we see the embedding of a scientific community within a larger society. Scientific communities have effects on technology, medicine, and education. Those communities are also affected by the economy and markets, by government policies, and attitudes to science in the culture as a whole.

Many interesting questions concern the relations between the three levels. We asked Popper, might an open-minded community be made up of closed-minded individuals? That is question about relations between levels 1 and 2. Much of the subtlety and interest in Kuhn's view is also about level 1 to level 2 relationships. Kuhn was not so interested in level 3; he was mostly concerned with scientific communities themselves. We will get to level 3 later.

Turning back to Kuhn, we have not yet reached the most controversial part of his theory, but are there any problems with what we have so far?

Kuhn probably exaggerates the degree of commitment that a normal scientist does and should have to a paradigm. Kuhn describes the attitude of a normal scientist in very strong terms. Scientific education is a kind of "indoctrination," which results in scientists having a deep "faith" in their paradigm. As a description of how science actually works, this seems exaggerated. Sometimes there is a faithlike commitment, but sometimes there is not. Many scientists are able to say that they work within a paradigm, for practical reasons, while being very aware of the possibility of error and the eventual replacement of their framework. One of the ironies of Kuhn's influence is that his book may have weakened the faith of some scientists, even though Kuhn thought that normal scientists should have a deep faith in their paradigms.

Leaving aside the factual issue of whether a tenacious commitment to a paradigm is what we generally find, we should also ask whether this strong commitment is a good thing. For Kuhn, the great virtue of normal science is its organized, coordinated structure. A constant questioning and criticism of basic beliefs is liable to result in chaos—in the partially random fact-gathering and speculation that we see in pre-paradigm science. But here again, Kuhn probably goes too far. He does not take seriously the possibility that scientists could agree to work together in a coordinated way, not wasting time on constant discussion of fundamental issues, while retaining a cautious attitude toward their paradigm. Surely this is possible.

5.5 *Revolutions and Their Aftermath*

"Look," Thomas Kuhn said. The word was weighted with weariness, as if Kuhn was resigned to the fact that I would misinterpret him, but he was still going to try—no doubt in vain—to make his point. "Look," he said again. He leaned his gangly frame and long face forward, and his big lower lip, which ordinarily curled up amiably at the corners, sagged. "For Christ's sake, if I had my choice of having written the book or not having written it, I would choose to have written it. But there have certainly been aspects involving considerable upset about the response to it." **John Horgan,** *The End of Science*

The most controversial parts of Kuhn's book were his discussions of scientific revolutions. Kuhn argued that some periods of scientific change involve a fundamentally different kind of process from what we find in normal science. Revolutionary periods see a breakdown of order and a questioning of the rules of the game, and they are followed by a process of rebuilding that creates new concepts, methods, and practices. Revolutions involve a breakdown, but they are essential to science as we know it. They have a "function," Kuhn often said, within the totality of science. The special features we associate with science arise from the combination and interaction of two different kinds of activity—the orderly, organized, disciplined process of normal science, and the periodic breakdowns of order found in revolutions. These two processes happen in sequence, within each scientific field. Science as a whole is a result of their interaction, and of nothing less.

Kuhn thought that looking *within* a period of normal science, you can easily distinguish good work from bad, rational moves from irrational, big problems from small problems, and so on. Progress is evident as time goes by. In a scientific revolution, as in a political one, rules break down and have to be rebuilt afresh. If you look at two pieces of scientific work across a revolutionary divide, it will not be clear whether there has been progress from earlier to later. It might not even be clear how to compare the theories or pieces of work at all—they may look like very different kinds of intellectual activity. The people on different sides of the divide will be "speaking different languages." In the climax of his book, Kuhn suggests that workers in different paradigms are living in different worlds.

How do revolutions occur? Above I described the transition from normal science to crisis. In Kuhn's story, large-scale scientific change usually requires both a crisis *and* the appearance of a new candidate paradigm. A crisis alone will not induce scientists to regard a large-scale theory or paradigm as "falsified." We do not find pure falsifications, rejections of one paradigm without simultaneous acceptance of a new one. Rather, the rejection of one paradigm accompanies the acceptance of another. But also, the switch to a new paradigm does not occur just because a new idea appears that looks better than the old one. Without a crisis, scientists will not have any motivation to consider radical change.

Suppose we do have a crisis, a period full of confusion and strange guests in the philosophy department. Then a new candidate paradigm appears, precipitating a revolution. What initially appears is a new paradigm in the narrow sense, an achievement that begins to inspire people and seems to point the way forward. More specifically, the new work appears to solve one or more of the problems that prompted the crisis in the old paradigm. The sudden appearance of problem-solving of this kind is the spark to the revolution. Kuhn did not think these processes could be described by an explicit philosophical theory of evidence and testing. Instead, we should think of the shift to a new paradigm as a something like a "conversion" phenomenon, or a gestalt switch. Kuhn also argued that revolutions are capricious, disorderly events. They are affected by idiosyncratic personal factors and accidents of history.

One reason for the disorderly character of revolutions is that some of the principles by which scientific evidence is assessed are themselves liable to be destabilized by a crisis, and they can change with a revolution. Kuhn did not argue that traditional philosophical ideas about how theories should relate to evidence are completely misguided. He made it clear in his later work that there are some core ways of assessing theories that are common to all paradigms (1977c, 321–22). Theories should be predictively accurate, consistent with well-established theories in neighboring fields, able to unify disparate phenomena, and fruitful of new ideas and discoveries. These principles, along with other similar ones, "provide *the* shared basis for theory choice" (322). (I should note that some commentators think these later essays change, rather than clarify, the views presented in *Structure*.) But Kuhn thought that when these principles are expressed in a broad enough way to be common across all of science, they will be so vague that they will be powerless to settle hard cases. Also, these goals must often be traded off against each other; emphasizing one often requires downplaying another.

Within a single paradigm, more precise ways of assessing hypotheses will operate. These will include sharper versions of the common principles listed above, but these sharper versions will not be explicit "principles." Instead, they will be more like habits and values, aspects of the shared mindset of normal scientists. Those are liable to change in the course of a revolution.

So we have two kinds of scientific change in Kuhn's picture, neither of which is what empiricist philosophies of science might have led us to expect. Change within normal science is orderly and responsive to evidence—but normal science works via a closing of debate about fundamental ideas. Revolutionary change does involve challenges to fundamentals, but these are episodes in which the orderly assessment of ideas breaks down. Displays of problem-solving power have an important role in these transitions between paradigms, but the shifts also involve gestalt switches and leaps of faith.

You might wonder at this point about the power of observational data to impose some order on these revolutionary changes. Kuhn, along with some others around the same time, argued that we cannot think of observation as a neutral source of information for choosing between theories, because what people see is influenced by their paradigm. This "theory-ladenness of observation" is an important topic in its own right, and it will be discussed in chapter 9.

In Kuhn's treatment of revolutionary change, the distinction between descriptive and normative issues is again important. Kuhn uses language that suggests that revolutions are not only bound to happen, but have a positive role in science. They are part of what makes science so powerful as a means for exploring the world (a "supremely efficient instrument"; 1996, 169). Some interpreters regard this as colorful talk and not essential to Kuhn's general message. I have the opposite view; I think this is central to Kuhn's picture. Science for Kuhn is a social mechanism that combines two capacities. One is the capacity for sustained, cooperative work. The other is science's capacity to partially break down and reconstitute itself from time to time. When a paradigm runs out of steam, there is nothing within the community that could reliably give science a set of directions for orderly movement toward a new paradigm. Instead, the goals of science are best served at these special times by a disorderly process, in which even basic ideas are put back on the table for discussion, and a new direction eventually emerges from the chaos. This sounds strange, but I think it was Kuhn's picture.

Here is another way of expressing these relationships. Without the tenacious commitment to a paradigm seen in normal science, investigation tends to be shallow. But without a descent into a crisis, there is no motivation to consider radically new ideas. Significant innovation requires both normal and revolutionary modes of change.

All of Kuhn's claims about what follows what in scientific change tend to be qualified; he is describing the central and characteristic patterns of change, not every case without exception. But the idea that revolutions generally require crises raised some serious historical issues. Was there a crisis in the state of astronomy before Copernicus, or in biology before Darwin? Was there a state of disorder following an earlier period of confident work? Maybe. But taking another biological example, if the appearance of genetics as a science around 1900 was a revolution, it is very hard to find a crisis in the work on inheritance that preceded it. Maybe Kuhn would regard this as a transition from pre-paradigm science to normal science, though that could not be said about most of biology around that time.

In the "Postscript" written for the second edition of *Structure*, Kuhn qualified his claims about the role of crisis (1970a, 181). He still maintained that crises are the "usual prelude" to revolutions. But even that claim is controversial. Kuhn's emphasis on crises sometimes seems driven more by the demands of his hypothesized mechanism for scientific change than by the historical data; Kuhn's story demands crises because only a crisis can loosen the grip of a paradigm and make people receptive to alternatives.

Another way there might be a revolution without a crisis comes from interactions between neighboring fields. Kuhn sometimes writes in a way that treats each scientific field as self-contained, but this is surely not so. Might there be a situation where there is a Kuhnian revolution in one field—with a crisis and all the rest—and the new paradigm that appears in that field also inspires a radical change in another field? This would be quite "un-Kuhnian" if things in the second field were previously going fairly well, but a revolution happens anyway. I don't have a clear example of this, but here is an approximate one. In the late 1950s there was a revolution in linguistics, owing to the work of Noam Chomsky (1957 and 1959). His "generative linguistics" introduced the idea that all humans have an innate knowledge of a grammatical "deep structure" that is common to all languages. For Chomsky, learning has a surprisingly minor role in how language develops in each of us. This might also be another case where you might wonder whether there was a crisis before the revolution, but in any case, once the new paradigm was established by Chomsky in linguistics, it had massive effects on a neighboring field, psychology. Earlier in this chapter I used as an example of normal science the behaviorist paradigm of Skinner in mid-twentieth-century psychology. This approach was replaced during and around the 1960s by "cognitive" psychology, using ideas of information processing, computation, and symbol manipulation. Several things fed into this shift (Greenwood 2015). Behaviorist views had problems, and new ideas surrounding the invention of computers seemed important for psychology. But the new approach that Chomsky introduced in the study of language was also inspiring, and a significant contributor to the revolution in psychology.

We might describe this case instead by saying that Chomsky was as much a psychologist as a linguist, and had effects on both fields. We might also wonder whether any of this fits Kuhn's model, in which there are paradigms that dominate a field until they die. But this case is a partial illustration of a revolution induced by a change in a neighboring field, and this does make sense as a possibility: a Kuhnian revolution in one field might prompt a no-crisis revolution in a neighboring field, simply by being an impressive and relevant breakthrough.

5.6 *Incommensurability, Relativism, and Progress*

Kuhn said that revolutions have a "non-cumulative" nature. There is no steady buildup of some useful outcome, like *true beliefs*, as science goes along. Instead, according to Kuhn, in a revolution you always gain some things and lose some things. Questions that the old paradigm answered now might become puzzling again, or cease to be coherent questions. So we might want to ask, do we usually gain more than we lose? In at least the middle chapters of his book, Kuhn seems to think there is no way to answer this question in an unbiased way (1996, 109, 110). Of course, it will *feel* like we have gained more than we've lost, or we would not have had the revolution at all. But that does not mean that there is some unbiased way of comparing what we had before with what we have after.

This question connects us to one of the most famous topics in Kuhn's work, the idea that different paradigms in a field are *incommensurable* with each other.

What does "incommensurable" mean here? Most literally, it means not comparable by use of a common standard or measure. This idea needs to be carefully expressed, however. Two rival paradigms can be compared well enough for it to be clear that they are incompatible—that they are rivals. And people working within any one paradigm will have no problem saying why their paradigm is superior to others, by citing key differences in what can be explained and what cannot. But these comparisons will be compelling only to those inside the paradigm from which the claim of superiority is being made. If we look "from above" at two people who are arguing during a revolutionary period, defending different approaches to their field, it will often appear that the two people are talking past each other.

There are two reasons for this—there are (roughly speaking) two aspects of the problem of incommensurability. First, people debating fundamental ideas will not be able to fully communicate with each other; they will use terms in different ways and in a sense will be speaking slightly different languages. Second, even when communication is possible, people in this situation will use different standards of evidence and argument. They will not agree on what a good theory is supposed to do.

Let us look first at the issues involving language. Here Kuhn's claims depend on a holistic view about the meaning of scientific language. Each term in a theory derives its meaning from its place in the whole theoretical structure. Two people operating within different paradigms might seem to use the same word—"mass" or "species"—but the meanings of these terms will be slightly different because of their different roles in the two rival theories.

For Kuhn, it will not usually happen that two people within different paradigms interact directly. As we saw, paradigms dominate and then are overthrown. But there are situations that are relevantly similar. The main one will involve one person defending an old paradigm and another person advocating a new one.

If incommensurability of meanings is real, then it should be visible in the history of science. Those who study the history of science should be able to find many examples of the usual signs of failed communication— confusion, correction, a perceived inability to make contact. Although I am not a historian of science, my impression is that historians have not found many examples of failed communication in crucial debates. Scientists are often adept at "scientific bilingualism," switching from one framework to another. And they are often able to improvise ways of bridging linguistic gaps, much as traders from different cultures are able to, by improvising "pidgin" languages (Galison 1997). Scientists often deliberately misrepresent each other's claims, in the service of rhetorical points, but that is not a case of failed comprehension or communication.

The other form of incommensurability is more important. This is incommensurability of *standards*. Kuhn argued that paradigms tend to bring with them their own standards for what counts as a good argument or good evidence, and these standards can change across a revolution.

One of Kuhn's best examples here involves the role of *causal explanation*. Should a scientific theory be required to make causal sense of why things happen? Should we always hope to understand the mechanisms underlying events? Or can a theory be acceptable if it gives a mathematical formalism that describes phenomena without making causal sense of them? An example of this problem concerns Newton's theory of gravity. Newton gave a mathematical description of gravity—his famous inverse-square law—but did not give a mechanism for how gravitational attraction works. Indeed, Newton's view that gravity acts instantaneously and at a distance seemed to be extremely hard to supplement with a mechanistic explanation. Was this a problem with Newton's theory, or should we drop the demand for a causal mechanism and be content with the mathematics? Would it be scientifically acceptable to regard gravity as just an "innate" power of matter that follows a mathematical law? People argued about this a good deal in the early eighteenth century. Kuhn's view is that there is no general answer to the question of whether scientific theories should give causal mechanisms for phenomena; this is the kind of goal that will be present in one paradigm and absent from another.

During the early twentieth century, there was a similar, although smaller-scale, debate within English biology. In the latter part of the nineteenth century, a group of biologists called the "Biometricians" had formulated a mathematical law that they thought described the inheritance of biological traits across generations (Provine 1971). They had no mechanism for how inheritance works, and their law did not lend itself to supplementation with such a mechanism. In 1900 the pioneering work done by Mendel in the mid-nineteenth century was brought to light, and the science of genetics was launched. For about six years, though, the Biometricians and the Mendelians conducted an intense debate about which approach to understanding inheritance was superior. One of the issues at stake was what kind of theory of inheritance should be the goal. The Biometricians thought that a mathematically formulated law was the right goal, while William Bateson, in the Mendelian camp, argued that understanding the mechanisms of inheritance was the goal. In the short term, the Mendelians won the battle. Eventually the two approaches were married; modern biology now has both math and mechanisms. But during the battle there was intense argument about what a good scientific theory should do (MacKenzie 1981). I agree with Kuhn that incommensurability of standards is a real and interesting issue.

Kuhn's discussion of incommensurability is the main reason his view of science is often referred to as "relativist." Kuhn's book is often considered one of the first steps in a tradition of work in the second half of the twentieth century that embraced relativism about science and knowledge. Kuhn himself was shocked to be interpreted this way. But what is relativism? This is a chaotic area of discussion. Roughly speaking, relativist views hold that the truth or justification of a claim, or the applicability of a rule or standard, depends on one's situation or point of view. Such a claim might be made generally ("all truth is relative") or in a more restricted way, about art, morality, good manners, or some other particular domain. The "point of view" might be that of an individual, a society, or some other group.

If people differ about the facts or the proper standards in some area, that itself does not imply that relativism applies in that case; some of the people might just be wrong. It is also important that if someone holds that moral rightness or good reasoning "depends on context," that need not be a form of relativism, although it might be. This is because a single set of moral rules (or rules of reasoning) might have built into them some sensitivity to circumstances. A set of moral rules might say, "If you are in circumstances X, you should do Y." That is not relativism, even though not everyone might be in circumstances X.

In this discussion we are mostly concerned with relativism applied to standards governing reasoning, evidence, and the justification of beliefs. And the "point of view" here is that of the users of a paradigm.

Is Kuhn a relativist with regard to these matters? Kuhn had a subtle view that is hard to categorize, and I doubt that everything Kuhn said can be fitted together consistently. The issue of relativism in Kuhn is also bound up with the question of how to understand scientific progress, something Kuhn struggled with in the final pages of his book.

As we have seen, Kuhn argued that different paradigms often carry with them different standards for good and bad scientific work. So far, this does not tell us whether Kuhn was a relativist—there might be an advance in standards as well in theories, where better ones replace worse ones as time passes. But Kuhn also argued that the paradigms we have in science now are not closer than earlier paradigms to an "ideal" or "perfect" paradigm. Scientific fields do not head steadily toward a final paradigm that is superior to all others.

Claims like these seem to be taking us close to a relativist view about the standards and ideas that are not shared across paradigms. But Kuhn said some rather different things in the final pages of *Structure*. There he said that our present paradigms have more problem-solving power than earlier paradigms did. This claim was made when Kuhn confronted the question of how to understand progress in science.

Kuhn gave two quite different kinds of explanation for the apparent large-scale progress we see in science. His first was a kind of eye-of-the-beholder explanation. Science will inevitably appear to exhibit progress because each field has one paradigm at a time, the victors after each revolution will naturally view their victory as progressive, and science is insulated from outside criticism. Celebrations of progress on the part

of the victors will not be met with any serious objection. This first explanation of the appearance of progress is consistent with a relativist view
of the changes between paradigms.

Kuhn also developed a different account of the appearance of progress
in science, especially in the final pages of *Structure*, and this account
seems to conflict with a relativist reading. Here Kuhn argued that science
has a special kind of efficiency, and this efficiency results in a genuine
form of progress across revolutions: the number and precision of solutions to problems in a scientific field tend to grow over time (1996, 170).
It is quite difficult to reconcile this claim with some of his discussions of
incommensurability in earlier chapters. There he said that revolutions
always involve losses as well as gains, and he also said that the standards
that might be used to classify some problems as important and others
as unimportant tend to change as a result of revolutions. It is not clear
whether the ideas about progress that Kuhn introduces in the last pages
of *Structure* are compatible with the rest of the book.

5.7 *The X-Rated "Chapter X"*

I have argued so far only that paradigms are constitutive of science. Now
I wish to display a sense in which they are constitutive of nature as well.
   Thomas Kuhn, *Structure*

Kuhn's book starts out with his analysis of normal science. The middle
chapters become more adventurous, and the book climaxes with chapter X. Here Kuhn puts forward his most radical claims. Not only do ideas,
standards, and ways of seeing change when paradigms change; in some
sense the *world* changes as well. Reality itself is paradigm-relative or
paradigm-dependent. After a revolution, "scientists work in a different
world" (1996, 135).

Philosophers and other commentators tend to split between two different attitudes toward this part of Kuhn's work. One group thinks that
Kuhn exposes the fact that any notion of a single, stable world persisting

through our various attempts to conceptualize it is an idea dependent on a failed view of science and outdated psychological theories. Kuhn, on this interpretation, shows that changing our view of science requires us to change our metaphysics too—our most basic views about reality and our relationship to it. Holding onto the idea of a single fixed world that science strives to describe is holding onto the last element of a conservative view of conceptual change.

Others think that this side of Kuhn's work is a mess. When paradigms change, ideas change. Standards change also, and maybe the way we experience the world changes as well. But that is very different from claiming that the world itself depends on paradigms. The way we see things changes, but the world itself does not change.

I am in the second camp; the X-rated chapter X is the worst material in Kuhn's great book. It would have been better if he had left this chapter in a taxi, in one of those famous mistakes that authors are prone to.

I should say immediately that it is not always clear how radical Kuhn wants to be. Sometimes it seems that he is just saying that our ideas and experience change. Also, there are some entirely reasonable claims we can make about changes to the world that result from paradigm changes. As paradigms change, scientists change their behavior and experimental practices as well as their ideas. Scientific revolutions result in new technologies that have far-reaching effects on the world. Probably many or most of the objects around you right now would not have existed at all if a lot of particular scientific theories had not been developed. Changes of this kind can be far-reaching, but they are still restricted by the causal powers of human action. We can change plants and animals by controlled breeding and genetic engineering. We can dam rivers and pollute them. We can create computers. But our reach is restricted, not indefinite. Kuhn discussed some cases in chapter X that make it clear that he did not have ordinary causal influences in mind. He discussed cases where changes in ideas about stars, planets, and comets led to astronomers "living in a different world," for example (1996, 117).

Perhaps the main problem with these discussions is that Kuhn seems to think that the view that we all inhabit a single world, existing independently of paradigms, also commits us to a naive set of ideas about perception and belief. This is an error. We might decide that perception is radically affected by beliefs and expectations, while still holding that perception is something that connects us to a single real world that we all inhabit.

Did Kuhn really make a mistake of this kind? Here is an especially relevant quote from the chapter:

> At the very least, as a result of discovering oxygen, Lavoisier saw nature differently. And in the absence of some recourse to that hypothetical fixed nature that he "saw differently," the principle of economy will urge us to say that after discovering oxygen Lavoisier worked in a different world. (1996, 118)

"Principle of economy"? Would it be economical for us to give up the idea that Lavoisier was living in the same world as the rest of us and acquiring new ideas about it? Appeals to economy are always suspicious in philosophy. They are usually weak arguments. This one also seems to have the accounting wrong.

From the point of view of a kind of skeptical philosophical discussion, it can be considered hypothetical that there is a world beyond our momentary sensory experiences and ideas. But this is a rather special sense of "hypothetical." If we are trying to understand science as a social activity, as Kuhn is, there is nothing hypothetical about the idea that science takes place in a single, structured world that includes the community of scientists, their instruments and laboratories, and various other objects, including the ones they try to study.

Kuhn hesitated after that quote above—maybe, he said, we should look into ways of "avoiding this strange locution." But he decided that we "must learn to make sense of statements that at least resemble these" (p. 121). I think Kuhn was not entirely sure what he wanted to say, but he was sure what he did *not* want to say. He did not want to say that although we might see the world differently and interpret it differently, we are all dealing with the same world, a world that might eventually change materially after a change of paradigm, as a result of new technologies and actions, but not just via the change in worldview itself. And that, I think, is what he should have said.

These issues connect to one raised in the previous section. Kuhn opposed the idea that the large-scale history of science involves an accumulation of more and more knowledge about how the world really works. He was willing, on occasion, to recognize some kinds of accumulation of useful results: there is an accumulation of problem-solving power. But we cannot see, in science, an ongoing growth of knowledge about the structure of the world.

When Kuhn wrote about this issue, he often came back to cases in the history of physics. Like Popper and others, Kuhn seems to have been hugely influenced by the fall of the Newtonian picture of the world at the start of the twentieth century. But Kuhn, along with some others, was too focused on the case of physics; he seems to have thought that we can only see science as achieving a growth of knowledge about the structure of the world if we can see this kind of progress in the parts of science that deal with the most low-level and fundamental entities and processes. If we look at other parts of science—at chemistry and molecular biology, for example—it is much more reasonable to see a continuing growth (with some hiccups) in knowledge about how the world works. We see a steady growth in knowledge about the structures of sugars, fats, proteins, and other important molecules, for example. There is no evidence that these kinds of results will come to be replaced, as opposed to extended, as science moves along. This type of work does not concern the most basic features of the universe, but it is undoubtedly science. I think that when we try to work out how to describe the growth of knowledge over time in science, we should probably treat theoretical physics as a special case and not as a model for all science (McMullin 1984). Kuhn's pessimism about the accumulation of knowledge in science appears overstated.

Although Kuhn's most famous discussions of realism are his notorious claims about how the world changes when paradigms change, at other times he seems more like a kind of pessimistic or skeptical realist. These are passages where Kuhn seems to think that the world is so complicated that our theories will always run into trouble in the end—and this is a fact about the world that is independent of paradigms. We try to "force" nature into "boxes," but nature resists. All paradigms are doomed to fail eventually, because nature is complex and science must simplify. This view is more coherent and more interesting than Kuhn's changing-worlds position.

5.8 *Final Thoughts about Kuhn*

Kuhn changed the philosophy of science by describing an extraordinarily vivid picture of scientific change. He attributed the success and the power of science to a delicate balance between factors in a complex and fragile mechanism; science owes its strength to an interaction between the ordered cooperation and single-mindedness of normal science, together with the ability of these behavioral patterns to break down and reconstitute themselves in revolutions. Quite quickly, critics were able to find problems with this mechanism when interpreted as a description of how science actually works—paradigms need not exert the kind of psychological dominance that Kuhn describes, and large-scale changes can occur without crises, for example. Many parts of Kuhn's mechanism are especially hard to apply to the history of biology. Kuhn's account of the mechanisms behind scientific change is in several ways too tightly structured, too specific. The real story is more mixed. But Kuhn's work was also an attempt at a new approach to the philosophy of science, a new kind of theory. These are theories that approach questions in the philosophy of science by looking at the social structure of science and the mechanisms underlying scientific change. This approach has flourished.

Back in the first chapter, I distinguished views that construe science broadly from those that construe it narrowly. Those that construe it broadly see the differences between science and everyday problem-solving as matters of detail and degree. Kuhn's theory is nothing like this. His theory of science emphasizes the differences between science and various other kinds of learning and investigation. Science is a form of organized behavior with a specific social structure. As a consequence, science appears in this story as a rather fragile cultural achievement; subtle changes in the education, incentive structure, and political situation of scientists could result in the loss of the special mechanisms of change that Kuhn described.

Before moving on, as a kind of appendix I will mention some connections between Kuhn's theory of science and a few other famous mechanisms for change. First, in some ways Kuhn's view of science has an "invisible hand" structure. The Scottish political and economic theorist

Adam Smith argued in *The Wealth of Nations* ([1776] 1976) that individual selfishness in economic behavior leads to good outcomes for society as a whole. The market is an efficient distributor of goods to everyone, even though the people involved are each just out for themselves. Here we have an apparent mismatch between individual-level characteristics and the characteristics of the group; selfishness at one level leads to the general benefit. The mismatch disappears when we look at the consequences of having a large number of individuals interacting together. (Smith's theory is another with interesting relations between individual-level and community-level facts.) Something similar is seen in Kuhn's theory of science: narrow-mindedness and dogmatism at the level of the individual lead to intellectual openness in science as a whole. Anomaly and crisis produce such stresses in the normal scientist that a wholesale openness to novelty is found in revolutions. In the next chapter we will look at a critic who was suspicious of Kuhn on exactly this point; he thought Kuhn was trying to excuse and encourage the most narrow-minded and unimaginative trends in modern science.

Another comparison requires a bit more background knowledge. In the chapter on Popper, I briefly compared his conjecture-and-refutation mechanism with a Darwinian mutation-and-selection mechanism in biology. A biological analogy can also be found in the case of Kuhn. During the 1970s the biologists Stephen Jay Gould, Niles Eldredge, and others argued that a large-scale pattern seen in much biological evolution is one of "punctuated equilibrium" (Eldredge and Gould 1972). A lineage of organisms in evolutionary time will usually exhibit long periods of stasis, during which we see low-level tinkering but little change to fundamental structures. These periods of stasis or equilibrium are punctuated by occasional periods of much more rapid change in which new fundamental structures arise. (Note that "rapid" here means taking place over thousands of years rather than millions.) The rapid periods of change are disorderly and unpredictable when compared to the simplest kind of natural selection in large populations. The periods of stasis also feature a kind of "homeostasis," in which the genetic system in the population tends to resist substantial change.

The analogy with Kuhn's theory of science is striking. We have the same long periods of stability and resistance to change, punctuated by unpredictable, rapid change to fundamentals.

The theory of punctuated equilibrium in biology was controversial for some time, especially because it was sometimes presented by Gould in rather radical forms (Gould 1980). The idea of a kind of homeostatic resistance to change brought about by the genetic system is a tendentious one, for example. And the idea that ordinary processes of natural selection do not operate normally during the periods of rapid change, but are replaced by other kinds of processes, is also very unorthodox. But as the years have passed, the idea of punctuated equilibrium has been moderated and has passed, in its more moderate form, into mainstream biology's description of some (not all) patterns in evolution.

Gould also wrote a paper called "Eternal Metaphors in Paleontology" (1977) in which he argued that the history of theorizing about the history of life sees the same basic kinds of ideas about change come up again and again, often mixed and matched into new combinations. The analogy between Kuhn's theory and the biological theory of punctuated equilibrium shows a similar kind of convergence in stories about processes of change. I say "convergence" here, but in some ways it's more than that. Gould (2002, 967) acknowledged the influence of Kuhn's picture of science on him when he was working out his biological ideas in the 1960s and 1970s.

Further Reading and Notes

Lakatos and Musgrave's *Criticism and the Growth of Knowledge* (1970) is an excellent set of essays on Kuhn. Another edited collection is Horwich's *World Changes* (1993). The fiftieth anniversary of Kuhn's book, in 2012, produced much reflection. See, for example, Devlin and Bokulich (2015).

Kuhn's collection of essays *The Essential Tension* (1977b) is an important additional source. Kuhn also wrote two historical books (1957; 1978). His later essays have been collected in *The Road since Structure* (2000).

For a detailed discussion of Kuhn's philosophy, see Hoyningen-Huene

(1993). Kitcher (1993) contains discussions of Kuhn's arguments about revolutions and progress. Doppelt (1978) is a clear discussion of incommensurability of standards.

The general question of whether there might be a revolution without a crisis, owing to another revolution in a neighboring field, was raised to me in discussion by Ramesh Ghelichi. Two important early works on the psychological side are Newell, Shaw, and Simon (1958) and Miller, Galanter, and Pribram (1960). Miller later said that encountering Chomsky at a seminar persuaded him to abandon behaviorism. According to Greenwood (2015), which I draw on here, many people involved in this episode in the 1960s saw it self-consciously in Kuhnian terms—they thought they were engaged in a Kuhnian revolution, and "everyone toted around their little copy of Kuhn's *The Structure of Scientific Revolutions*" (James Jenkins, quoted in Greenwood 2015, 454).

Chapter 6

Theories and Frameworks

6.1 *After* Structure

The period after the publication of Kuhn's book was one of intense and sometimes heated discussion in all the fields that try to understand science. In this chapter we will look at some other philosophical accounts of science developed around this time, all of them developed in interaction with Kuhn or in response to him. We'll also look at some general philosophical themes coming out of the last few chapters.

First, we will look at the views of Imre Lakatos. Lakatos's main contribution was the idea of a *research program*. A research program is similar to a paradigm in Kuhn's (broad) sense, but it has a key difference: we expect to find more than one research program in a scientific field at any given time. The large-scale processes of scientific change can be understood as competition between research programs.

It should be clear from the previous chapter that this was an idea waiting to be developed. Kuhn's insistence that scientific fields usually have only one paradigm operating at any time was criticized from the initial publication of *Structure*. Lakatos was the first person to develop a picture of science in which larger paradigm-like units operate in parallel and compete in an ongoing way. Lakatos's own development of this idea does have problems. Soon after, another philosopher, Larry Laudan, worked out a better version of the same basic idea.

Then we turn to Paul Feyerabend, the wild man of twentieth-century philosophy of science. Feyerabend is probably the most controversial figure contributing to the debates discussed in this book. I called him "the" wild man, though there have been various other wild men and women in the field besides Feyerabend. But Feyerabend's voice was uniquely wild. He argued for "epistemological anarchism," a view in which rules of method and normal scientific behavior were to be replaced by a free-wheeling attitude in which *anything goes*.

Kuhn, Lakatos, and Feyerabend all interacted and developed some of their ideas in response to each other (with the exception perhaps of much influence of Lakatos on Kuhn). Feyerabend's most important work (1975) was written, he said, as a kind of letter to Lakatos, but Lakatos died in 1974 before writing a reply.

6.2 *Lakatos and Research Programs*

Imre Lakatos was born in Hungary, became a communist, and was a member of the resistance to Nazi occupation during World War II. After the war he worked in politics and was jailed for over three years by the Stalinist regime. He left Hungary, made his way to England, and eventually ended up at the London School of Economics, working with Popper. Lakatos often claimed that his main ideas about science were implicit in Popper or were one side of Popper's views. Although there is some truth in this, it is better to consider Lakatos's ideas on their own terms.

Lakatos saw Kuhn's influence as *destructive*—destructive of reason and dangerous to society. For Lakatos, Kuhn had presented scientific change as an irrational process, a matter of "mob psychology" (1970, 178), where the loudest, most energetic, and most numerous voices would prevail. The interpretation I gave of Kuhn in the previous two chapters is different; I think Kuhn saw science as an almost miraculously well-structured machine for exploring the world. Even the disordered episodes found in revolutions have a positive role in the functioning of the whole. Lakatos, in contrast, saw the disorder in Kuhn's picture as no more than dangerous chaos. Lakatos wanted to rescue the rationality of science from the damage Kuhn had done.

Lakatos called his view a *methodology of scientific research programs* (1970). A research program is similar in some ways to a Kuhnian paradigm (in the broad sense), but there is usually more than one research program operating in a field at any time. For Lakatos, competition between research programs is what we actually find in science, and it is also essential to rationality and progress. This view was applied to all of science, from physics to the social sciences. A research program will evolve over time; it will contain a sequence of related theories. Later theories are developed in response to problems with the earlier ones. For Lakatos, as for Kuhn, it is common and justifiable for a research program to survive for a while despite empirical anomalies and other problems. Workers within a research program typically have some commitment to the program; they do not reject the basic ideas of the program as soon as something goes wrong. Rather, they try to modify their theories to deal with the problem. However, for Lakatos, as for Kuhn, research programs are sometimes abandoned. So a complete theory of scientific change must consider two different kinds of change: (1) change within individual research programs, and (2) change at the level of the collection of research programs within a scientific field.

A research program has two main components, in Lakatos's view. First, it contains a *hard core*. This is a set of basic ideas that are essential to the research program. Second, a research program contains a *protective belt*. This is a set of less fundamental ideas that are used to apply the hard core to actual phenomena. The detailed, specific versions of a scientific theory that can actually be tested will contain ideas from the hard core combined with ideas from the protective belt.

For example, the Newtonian research program of eighteenth-century physics has Newton's three laws of motion and his gravitational law as its hard core. The protective belt of Newtonianism will change with time, and at any time it will contain ideas about matter, a view about the structure of the universe, and mathematical tools used to link the hard core to real phenomena. The nineteenth-century Darwinian research program in biology has a hard core that claims that different biological species are linked by descent and form a family tree (or perhaps a very small number of separate trees). Changes in biological species are due mostly to the accumulation of tiny variations favored by natural selection, with some other causes of evolution playing a secondary role. The protective belt of nineteenth-century Darwinism is made up of a shifting set of more detailed ideas about which species are closely related to which, along with ideas about the workings of inheritance, variation, competition, and natural selection.

We now reach Lakatos's principles of scientific change. Let us first look at change within research programs. The first rule is that changes should be made only to the protective belt, never to the hard core. The second rule is that changes to the protective belt should be "progressive." Here Lakatos borrowed from Popper's ideas. A progressive research program constantly expands its application to a larger and larger set of cases, or strives for a more precise treatment of the cases it presently covers. A progressive research program is one that is succeeding in increasing its predictive power. In contrast, a research program is "degenerating" if the changes that are being made to it only serve to cover existing problems and do not successfully extend the research program to new cases. Lakatos assumed, like Kuhn, that all research programs are faced with anomalies—unsolved problems—at any time. A degenerating research program is one that is falling behind, or only barely keeping up, in its attempt to deal with anomalies. (Lakatos's term "degenerating research program" is a useful one that quickly caught on and is now seen in a variety of contexts.) A progressive research program, in contrast, fends off refutation but also extends itself to cover new phenomena. Lakatos thought that, in principle, we could measure how quickly a research program is progressing.

Now let us look at the other level of change in Lakatos's system, change at the level of the collection of research programs present in a scientific field.

Each field will have a collection of research programs at any given time, some of which are progressing rapidly, others progressing slowly, and others degenerating. You might be thinking that the next rule for Lakatos is obvious: "choose the most progressive research program." That would establish a usable procedure for scientists looking at their field as a whole, working out what to do, and it would give us a way of deciding who is making rational or irrational decisions. But that is not what Lakatos said.

For Lakatos, it is acceptable to protect a research program for a while during a period when it is degenerating—the research program might recover. This is even the case when another research program has overtaken it (Lakatos 1971). The history of science contains cases of research programs recovering from temporary bad periods. So a reasonable person can wait around and hope for a recovery. How long is it reasonable to wait? Lakatos does not say. Feyerabend, who I will discuss later in this chapter, swooped on this point (1975). For him it was the Achilles' heel in Lakatos's whole story. If Lakatos does not give us a rule for when a rational scientist should give up on one research program and switch to another, his account of the rational choice of theories is empty.

So is there a third rule that tells us how to handle decisions between research programs? Not really. Lakatos did say that the decision to stay with a degenerating research program is a high-risk one (1971). So Lakatos might advise the rational scientist to stay with a degenerating research program only if he or she is willing to tolerate a high-risk situation. And Lakatos is right that different people can reasonably have different attitudes toward risk. But Lakatos did not follow the suggestion up, and he left this whole matter quite incomplete. The tremendous *appearance* of order and strictness in Lakatos's philosophy of science is largely undermined by his failure to say something definite about this point. Feyerabend was right to see a mismatch between the rhetoric and the reality of Lakatos's views. Indeed, it sometimes looks as if the whole point of Lakatos's project was just to give us a way of retrospectively describing episodes in science as rational.

This is related to other odd features in Lakatos's work. He wrote about science in a moralistic way, always talking about "duties" and so on, but he also often wrote in exaggerations, with tongue in cheek, even introducing fantasies that warp the historical record (see Lakatos 1970, 138 n., 140 n.). Here he comes across as something of a trickster figure. I find him quite puzzling.

This is also a good place to emphasize the difference between Lakatos and Kuhn in their underlying attitudes. Kuhn has a deep trust in the shared standards implicit in paradigms and the ability of science to find a way forward after crises, despite some groping and flailing along the way. For Kuhn, once we rid our picture of science of some myths, the picture we are left with is fundamentally healthy. Kuhn trusts science left in the hands of implicit shared values. Lakatos, in contrast, wants to have the whole enterprise guided by methodological rules—or at least, he needs for us to be able to tell ourselves a story of that kind.

For all these oddities, the basic idea of competing research programs is certainly a useful one. There are some fields and episodes where this seems a far more accurate description than Kuhn's paradigm-based view. Present-day psychology is an example. We can also consider the possibility of *mixtures* of Kuhn-like and Lakatos-like stories. In biology, what we often find is consensus about very basic principles but competition between research programs at a slightly lower level. Looked at very broadly, evolutionary biology might contain something close to a single paradigm: the "synthetic theory," a combination of Darwinism and genetics. But at a lower level of generality, we seem to find competing research programs. The "neutral theory of molecular evolution" is a research program that tries to understand most variation and change at the molecular genetic level in terms of random processes rather than natural selection (Kimura 1983). This research program is compatible with the central claims of the synthetic theory, but it conflicts with some standard ways in which the synthetic theory is applied to variation within populations. Another research program of this kind is the "evo-devo" movement (the name is short for "evolutionary developmental biology"), which seeks to understand various kinds of evolutionary change in terms of changes in the mechanisms that take an organism along a path from a juvenile to adult form (Hall 2003).

In biology and elsewhere, often a new research program is "budded off" from the mainstream of the field and explored for a few decades to see how much it can explain. If it does well, it may contribute ideas that filter back into mainstream thinking, and it fades once it stops contributing these new ideas.

We now have tools for describing a range of different kinds of scientific change. Some fields may have dominant paradigms and Kuhnian normal science. Others may have competing research programs. These might also be seen as two poles, where a field can move along a line from one to the other:

Dominating paradigm ◄———————► Competing research programs

Some fields might have very general paradigms plus various lower-level research programs budding off periodically. (Here I should note that the term "research program" can also be used to describe different noncompeting approaches within a single field.)

I have been discussing the usefulness of the research program idea in describing how science actually works. There is also the possibility of *normative* theories that make use of this concept. But I will not follow up that idea further within Lakatos's framework; I'll return to the topic after introducing a related picture.

6.3 *Laudan and Research Traditions*

In a book called *Progress and Its Problems* (1977), Larry Laudan developed a view that is similar to Lakatos's in basic structure but is superior in a number of ways. Like Lakatos, Laudan thought that Kuhn had described science as an irrational process, as a process in which scientific decision-making is "basically a political and propagandistic affair" (1977, 4). This reading of Kuhn (I say yet again) is inaccurate. But Laudan also recognized the power of Kuhn's discussions of historical cases. Like Lakatos, Laudan wanted to develop a view in which paradigm-like units could coexist and compete in a scientific field. He gave many cases from the history of science to motivate this picture. So we are heading toward the idea of a research program. But in an understandable piece of product differentiation, Laudan called the large-scale units of scientific work "research traditions" rather than research programs.

The difference between Laudan and Lakatos is not just terminological. Lakatos saw the sequence of theories within a research program as very tightly linked: each new theory was supposed to have a broader domain of application than its predecessor in that research program, and for Lakatos, the hard core never changes. For Laudan, the theories grouped within research traditions are more loosely related. There can be some movement of ideas in and out of the hard core. Moreover, for Laudan there is nothing unusual or bad about a later theory covering less territory than an earlier one; sometimes a retreat is necessary. In Laudan's picture, theories can also break away from one research tradition and be absorbed by others. For example, the early thermodynamic ideas of Sadi Carnot were developed within a research tradition that saw heat as a fluid ("caloric"), but these ideas were taken over in time by a rival research tradition that saw heat as the motion of matter.

Another innovation in Laudan's account is his distinction between the *acceptance* and the *pursuit* of theories.

The philosophies discussed in this book so far have tended to recognize just *one kind* of attitude that scientists can have to theories. Usually, the attitude of a scientist to a theory has been treated as something like belief. Belief can come in degrees, it seems; there are cautiously held and firmly held beliefs. But so far there seems to be one basic kind of attitude—belief or something like it. Laudan argued that there are two different kinds of attitudes to theories and research traditions found in science: acceptance and pursuit. Acceptance, for Laudan, is close to belief; to accept something is to treat it as true. But pursuit is different. It involves deciding to work with an idea and explore it, and this might be done for reasons other than confidence that the idea is likely to be true. It can be reasonable to pursue an idea that one definitely does not accept. Someone might think that *if* some adventurous new idea were true, it would be of huge importance and the payoff would be high. Someone else might think that although some idea is not likely to be true, it should be explored, and she or he is the person best equipped to do so. There is a whole constellation of different reasons that a person might have for working with a scientific idea.

Using this distinction, Laudan was able to give some fairly sharp rules of choice where Lakatos had not. For Laudan, it is always rational to pursue the research tradition that has the highest current rate of progress in problem-solving (1977, 111). But that does not mean one should *accept* the basic ideas of that research tradition. The acceptability of theories and ideas is measured by their present overall level of problem-solving power, not by the rate of change. We should accept (perhaps cautiously) the theories that have the highest level of problem-solving power. So a scientist might be inclined to accept the ideas in a mainstream research tradition but work on a more marginal research tradition that has a spectacular rate of progress. For Laudan that decision would be a rational one.

With any rule like this, it will be possible to think of cases where the rule might lead a person astray. What if a research tradition has a low rate of progress right now, but there is good reason to think it might take off very soon? This is the kind of possibility that made Lakatos hesitate.

Laudan clearly hoped that the distinction between acceptance and pursuit would help with this kind of problem, and so it does. But he does not try to lay down rules that will deal with every possible situation, including all the various kinds of bad luck. And when we think about these problems, it can be quite unclear what kinds of principles a philosophy of science should be looking for. Some people think that looking for general rules of procedure, even sophisticated ones like Laudan's, is just a mistake—good decisions always tend to depend on specific features of particular cases. But it is fair to say that Laudan was able to give quite an impressive normative theory of science using the idea of competition between research traditions.

Laudan's theory was impressive, but here is an interesting gap in both theories discussed in this chapter that some readers may have picked up on already. Both Lakatos and Laudan were interested in the situation where a scientist is looking out over a range of research programs in a field and deciding which one to join. But here is a question that neither of them seemed to give much thought to: does the answer depend on how many people are already working in a given research program? Both Lakatos and Laudan seemed to think that it would be fine for their theories to direct everyone to work on the same research program, if it was far superior to the others. But perhaps that is a mistake. Science might be better served by some mechanism in which the field "hedges its bets." That suggests a new question: what is the best distribution of workers across a range of research programs?

There are two different ways of approaching this question. One way is to look at individual choices. Does it make sense for *me* to work on research program 1 rather than research program 2, given the way people are already distributed across the two programs? Is research program 1 overcrowded? We can also approach the issue another way. We can ask: which distribution of people across rival research programs is best for science?

Kuhn was aware of this issue, especially in his work after *Structure* (e.g., Kuhn 1970b). This is ironic because Kuhn did not think that ongoing competition between paradigms was usually found in science. But Kuhn did say that one of the strengths of science lay in its ability to distribute risk by having different scientists make different choices, especially during crises. Lakatos and Laudan were in a good position to make a thorough investigation of this issue, but they did not (see also Musgrave 1976). It was not until more recently that this question was brought into sharp focus. I will take up the issue again in chapter 9.

6.4 *Anything Goes*

Now we turn to Paul Feyerabend, the most controversial figure in the post-Kuhn debates. Feyerabend, like a number of important people in this book, was born in Austria. He fought in the German infantry during World War II. He switched from science to philosophy after the war and eventually made his way to the University of California, Berkeley, where he taught for most of his career. Feyerabend was initially influenced by Popper, but by the early 1960s he was moving toward the adventurous views for which he became famous.

Feyerabend is not as influential now as he once was, but he had an important role in the development of ideas about science. He was bold and explicit in developing a radical position based on the collapse of logical empiricist ideas—he showed how radical the shift from those ideas could be. (He was also a much better writer than most radicals have been since then.)

What were Feyerabend's notorious ideas? A two-word summary gets us started: anything goes. In his book *Against Method* (1975), Feyerabend argued for "epistemological anarchism." The epistemological anarchist is opposed to all systems of rules and constraints in science. Great scientists are opportunistic and creative, willing to make use of any available technique for discovery and persuasion. All attempts to establish rules of method in science will result only in straitjacketing this creativity. We see this, Feyerabend said, when we look at the history. Great scientists have always been willing to break even the most basic methodological rules that philosophers, and other people obsessed with order, might try to lay down.

Like Kuhn, Feyerabend thought that rival scientific theories are often incommensurable (see section 5.3 above; Feyerabend and Kuhn developed these ideas in tandem). He argued that observations in science are contaminated with theoretical assumptions and hence cannot be considered a neutral test of theory. Some of these arguments were based on speculative ideas about scientific language, and are not very convincing. His more interesting arguments are of two kinds. These involve the history of science and also a direct confrontation of some hard questions about how science relates to freedom and human well-being.

Before launching into this unruly collection of ideas, we need to keep in mind a warning that Feyerabend gave at the start of *Against Method*. He said that the reader should not interpret the arguments in the book as expressing Feyerabend's "deep convictions." Instead, they "show how easy it is to lead people by the nose in a rational way" (1975, 32). The epistemological anarchist is like an undercover agent who uses reason in order to destabilize it.

It is hard to know what to make of this, but I think it is possible to sort through Feyerabend's claims and distinguish some that do represent his deep convictions. Feyerabend's deepest conviction was that *science is an aspect of human creativity*. Scientific ideas and scientific change are to be assessed in those terms.

Michael Williams (1998) has suggested that we think of Feyerabend as a late representative of an old skeptical tradition, represented by Sextus Empiricus and Montaigne, in which the skeptic "explores and counterposes all manner of competing ideas without regarding any as definitely established." This is a useful comparison, but it is only part of the story. To capture the other part, we might compare Feyerabend to Oscar Wilde, the nineteenth-century Irish playwright, novelist, and poet who was imprisoned in England for homosexual behavior and never recovered from this ordeal. Wilde liked to express strange, paradoxical claims about knowledge and ideas ("I can believe anything so long as it's incredible"). But behind the paradoxes there was a definite message. For Wilde, the most important kind of assessment of ideas is aesthetic assessment. A book or an idea might look immoral or blasphemous, but if it is beautiful, then it is worthwhile. Other standards—moral, religious, logical—should never be allowed to get in the way of the free development of art.

This, I suggest, is close to Feyerabend's view; what is important in all intellectual work, including science, is the free development of creativity and imagination. Nothing should be allowed to interfere with this.

Feyerabend's focus on values and creativity guided his readings of others. His paper "Consolations for the Specialist" (1970) shows him to be one of the most perceptive critics of Kuhn. Most philosophers found disorder in Kuhn's view of science. Feyerabend found the opposite: an incitement for scientists to become orderly and mechanical. Feyerabend saw Kuhn as glorifying the mind-numbing routine of normal science and the rigid education that Kuhn thought produced a good normal scientist. He saw Kuhn as encouraging the worst trends in twentieth-century science toward professionalization, narrow-mindedness, and exclusion of unorthodox ideas.

Feyerabend recognized the invisible-hand side of Kuhn's story, his attempt to argue that individual narrow-mindedness is all for the best in science. Back in chapter 5 I said that it is often hard to distinguish the descriptive from the normative in Kuhn's discussions. Feyerabend saw this ambiguity in Kuhn's writing as a deliberate rhetorical device for insinuating into the reader a positive picture of the most mundane type of science. For Feyerabend, the mindset that Kuhn encouraged also leads to a lack of concern for the moral consequences of scientific work.

In addition, Feyerabend argued that Kuhn was factually wrong about the role of normal science in history. According to Feyerabend, paradigms almost never succeed in exerting the kind of control Kuhn described. There are always imaginative individuals trying out new ideas.

Feyerabend was not, as he is sometimes portrayed, an "enemy of science." He was an enemy of *some kinds* of science. In the seventeenth century, according to Feyerabend, science was the friend of freedom and creativity and was heroically opposed to the stultifying grip of the Catholic Church. He admired the scientific adventurers of this period, especially Galileo. But the science of Galileo is not the science of today. Science, for Feyerabend, has gone from being an ally of freedom to being an enemy. Scientists are turning into "human ants," unable to think outside of their training (1975, 188). And the dominance of science in society threatens to turn man into a "miserable, unfriendly, self-righteous mechanism without charm or humour" (175). In the closing pages of *Against Method*, he declares that society now has to be freed from the strangling hold of a domineering scientific establishment, just as it once had to be freed from the grip of the One True Religion.

6.5 *An Argument from History That Haunts Philosophy*

A large part of Feyerabend's most famous book, *Against Method*, is taken up with a discussion of Galileo Galilei's arguments against his Aristotelian opponents in the early seventeenth century. Galileo aimed to defend the literal truth of Copernicus's claim that the Earth goes round the sun rather than vice versa. One of the things Galileo had to confront was a set of obvious arguments from experience against a moving Earth. For example, when a ball is dropped from a tower, it lands at the foot of the tower, even though, on Copernicus's view, the tower has moved a significant distance (along a huge circle) while the ball is in the air. All of our everyday experience of motion suggests that the Earth is stationary. These are not arguments from the wisdom of Aristotle or the sayings of the biblical apostles; they are arguments from what we observe every day. If empiricism in philosophy has any teeth at all, Feyerabend claims, it entails that people in the seventeenth century had excellent reasons to resist Galileo and believe that the Earth did not move.

Galileo rejected the arguments. In his *Dialogue Concerning the Two Chief World Systems* ([1632] 1967), he patiently tries to show that Copernicus's model is compatible with everyday experiences of motion. If the Earth is moving, then a ball dropped from a tower has a *mixed* kind of motion. It is falling toward the Earth, but is also moving in a huge circle, just as the tower is. Our everyday perception of motion is unable to distinguish the case where both tower and ball have a circular motion from the case where neither does.

Galileo defuses the arguments, but he does not suggest that this is easy. Indeed, he marvels at Copernicus and others who have "through sheer force of intellect done such violence to their own senses as to prefer what reason told them over that which sensible experience plainly showed them to be the contrary" (quoted in Feyerabend 1975, 101).

According to Feyerabend, what Galileo had to do was to create a different kind of observational description of the world, one in which descriptions of apparent motion were compatible with the Copernican hypothesis. Only then would the arguments for Copernicanism become plausible. What science had to do, here and in other cases, was to break through the constraints of an outdated worldview that permeated even the most basic observational description. Science, for Feyerabend, is often a matter of challenging rather than following the lessons of observation.

This is an example of something that Feyerabend regarded as ubiquitous. A basic empiricist principle, of the kind dear to philosophers, would seem to be pointing people in the seventeenth century away from the scientific theory that we think now to be true. The philosopher complacently spouts generalities about how science is powerful because it is responsive to observational data. But history suggests that the principles the philosopher likes so much would steer us in the wrong direction if people back in the crucial period had applied them.

Through all the exaggerations, provocations, jokes, and outrageous statements in Feyerabend's works, this form of argument runs as a constant and challenging thread. Are there any principles of method, measures of confirmation, or summaries of the scientific strategy that do not fail the great test of the early seventeenth century? Look at the massiveness of the rethinking that Galileo urged, and the great weight of ordinary experience telling against him. Given these, would all traditional philosophical accounts of how science works, especially empiricist accounts, have instructed us to stick with the Aristotelians rather than take a bet on Galileo? This is the Feyerabendian argument that haunts philosophy of science.

Feyerabend massively overextends his argument, though, into a principle that cannot be defended: "Hence it is advisable to let one's inclinations go against reason *in any circumstances*, for science may profit from it" (1975, 156). Feyerabend claims that because some principle or rule *may* go wrong, we should completely ignore it. The claim is absurd. The policy of catching trains that are scheduled to take you where you plan to go is a policy that may go wrong. The train might crash. Or if you caught a different train, you might meet the love of your life on the way. All that is possible, but no rational person regards these mere possibilities as sufficient to discredit the rule that it is best to take trains that are scheduled to go where you want to go. All applications of everyday rules of rational behavior presuppose judgments about which outcomes are typical or probable, and which are far-fetched or unlikely. In rational behavior, nothing is guaranteed, but some policies and rules can be justified despite this.

What is true for everyday behavior is true also for science. The mere possibility that a rule might lead to bad consequences proves very little. We need more than a mere possibility before we have grounds to doubt a principle. Science "may profit" from all kinds of strange decisions, but "may" is not enough.

So we find a mixture of good and bad argument in Feyerabend's treatment of these issues. It would be foolish for a philosopher to ignore the strangeness of having one's favorite principles of theory choice come out in favor of the seventeenth-century Aristotelians and against Galileo. The example is so important in the history of science that it takes a brave philosopher to ignore it. It is by no means clear that well-developed philosophical theories do make the wrong call, though. Recall Laudan's rules discussed in section 6.3 above; they would probably steer us rather well here. These rules would at least tell us to *pursue* the Galilean program, on account of the dramatic way it was increasing its problem-solving power. And in time, it would become rational to accept the Galilean view.

6.6 *Frameworks, Theories, and Empiricism*

In this section I take stock, having now covered a series of philosophical debates that run from the early twentieth century to the end of its third quarter. We started out with empiricism, a family of attempts to give a theory of knowledge, including scientific knowledge, in terms of the guidance of experience. Early twentieth-century versions of empiricism ran into trouble of several kinds; the chapter on evidence and induction (chapter 3) was a chapter of frustrations and surprises. Popper developed a view that was empiricist in some ways but rejected induction and made more of the creative, imaginative side of science. This role for the imagination had not been denied by the empiricist tradition, but it often seemed to be well in the background. For Popper, bold creative leaps are essential to science. Kuhn put a different kind of pressure on empiricist ideas, especially through his views about paradigms, organizing structures that enable empirical work to be done effectively, and are resistant to change until the pressures become extreme.

The role of *organizing frameworks* of various kinds is a recurring theme in this area, very powerful in Kuhn's work but also seen in other forms. This included a role in the later stages of logical empiricism. Carnap, in his late work, argued that we deal with experience with the aid of "linguistic frameworks" that are used to categorize events and organize things. Many years earlier, Immanuel Kant was perhaps the first philosopher to develop a detailed view in which an abstract conceptual framework acts to guide empirical investigation ([1781] 1998). According to Kant, the basic framework that we bring to bear on the world is fixed and universal across all normal humans. We cannot, as humans, escape this framework (and would not want to if we could). Frameworks of the sort Carnap had in mind are different from Kant's, as we can freely choose to adopt or reject them. Treating the world as containing physical objects is one such framework (perhaps it's not so clear we could reject this one). Another is the framework of mathematics, in which numbers, sets, and so on are treated as real things we can study and describe (Carnap 1956).

Carnap distinguished moves made *within* these frameworks from changes made *between* them. The principles that are fundamental to a framework will appear as analytic sentences (see section 2.3) if they are explicitly stated. Moves made within a framework involve the assessment and testing of synthetic claims. Switches between frameworks involve a different kind of process from moves made within them. A framework cannot be falsified in the way that claims about particular events or objects (or numbers) can be. Moves made between frameworks are based on a kind of pragmatic assessment of the overall usefulness of the framework. If one framework seems not to be working, then we try another.

Quine, in "Two Dogmas of Empiricism" (1951a) and elsewhere, argued against this two-layered view of language and knowledge. For him, all changes made to our belief system, whether large or small, involve the same kind of holistic tinkering with the web of belief. We accommodate experience by making as few changes as possible and keeping our worldview as simple as we can. Quine rejected any distinction between changes within and changes between frameworks. There is just change.

Kuhn can be seen as similar to Carnap in some ways here, and hence different from Quine. Kuhn's paradigms are much richer than Carnap's linguistic frameworks; they are not just ways of talking and categorizing, but whole ways of doing science. But there is again a distinction here between moves *within* paradigms/frameworks and moves *between* them. Later in the century other views of this general kind appeared—we looked at a couple in this chapter. The idea of a framework also provides a road to the development of relativist views of various kinds. Perhaps justification, and even truth, only exist in a way that is internal to a framework.

Feyerabend recognized the psychological power of linguistic and cosmological frameworks, but he insisted that the imaginative person can resist the bounds of a framework. Popper also rejected the whole idea of frameworks as constraining thought and knowledge. He called it "the myth of the framework" (1976). Feyerabend did not see frameworks as mythical, but he thought that their bounds could be resisted and overcome.

Another interesting response to these issues is seen in Peter Galison's work (1997). Galison argues that big changes in the different elements making up a scientific discipline are not in step with each other. Whereas Kuhn described a process in which there is simultaneous change in theoretical ideas, methods, standards, and observational data—all working as a package—Galison argues that within physics, change in experimental traditions does not tend to occur simultaneously with a big change in theory. This is because of the partial autonomy of these different aspects of large-scale science. (Instrumentation is yet another tradition, for Galison, with its own rates and causes of change.) A challenging theoretical shift will be made more manageable by the fact that we can expect other aspects of the same field not to be changing at the same time. Disruptions happen more locally than they do in the Kuhnian model, and there are more resources available to the field to negotiate the transitions in an orderly way. The history of a scientific field shows lots of "seams" or breaks, but the seams on the side of theory and the seams on the side of experimental work do not line up. Galison thinks that there is a sort of strength that scientific traditions acquire from this structure, much as a brick wall is stronger when the seams or divides between the bricks in different layers are offset from each other, creating a lot of overlap between them.

Returning to the historical story, by the third quarter of the twentieth century most philosophers who called themselves "empiricists" would probably have seen their view as under siege. Holistic versions of empiricism, with something like Quine's web-of-belief picture, were the most common. The empiricist philosophical tradition also seemed, in these years, to have missed the boat with respect to the social side of science and knowledge. It tended to be set up in a rather individualist way—not denying the social side, but not doing much with it. There were exceptions to this; the American philosopher John Dewey regarded himself as an empiricist, though an unorthodox one, and his view of knowledge and science placed great emphasis on communities and social interaction. Dewey was part of the "pragmatist" tradition in philosophy, working mostly in the earlier part of the century. C. S. Peirce, who began that pragmatist lineage in the late nineteenth century, also had an epistemological view based on social processes, especially on the way scientific communities iron out errors over time and achieve consensus (see section 13.2). Those ideas were some distance from the mainstream in philosophy of science through the period I've been writing about, though. They began to get more recognition later.

Further Reading and Notes

Lakatos's most famous work is his long paper in Lakatos and Musgrave's *Criticism and the Growth of Knowledge* (1970). Another key paper is

Lakatos (1971). Cohen, Feyerabend, and Wartofsky (1976) is a collection of his work. Lakatos also did influential work in the philosophy of mathematics (1976). Lakatos's eccentric personality has often been regarded with a kind of amused affection, but historical work has uncovered a much darker side to his early years: see Musgrave and Pigden (2016).

Feyerabend's most important book is *Against Method* (1975), but his earlier works, collected in Feyerabend (1981), are also significant. His later books are not as good, though *Science in a Free Society* (1978) has some interesting parts. *The Worst Enemy of Science?* (Preston, Munévar, and Lamb 2000) is a collection of essays on Feyerabend. Horgan (1996) contains another great interview.

The relationship between Lakatos and Feyerabend is documented in detail in Motterlini (1999). Carnap's most famous article on frameworks is his "Empiricism, Semantics, and Ontology" (1956), which is very readable, by Carnap standards.

Peirce's "The Fixation of Belief" (1877) is an important early statement of a social view of scientific inquiry. *The Quest for Certainty* (1929b) is the best introduction to Dewey's thinking.

Chapter 7

The Challenge from Sociology of Science

7.1 Beyond Philosophy?

As the twentieth century moved further into its second half, thinking about science became more diverse and increasingly radical options appeared. The previous chapter looked at some of these developments from the side of philosophy. The same phenomenon is found—perhaps even more so—in neighboring fields. I will focus on sociology of science, an area that began to interact intensely with philosophy. Some of the same issues arose in the history of science, but it was sociology that sometimes set itself up as a replacement, or "successor discipline," to the philosophy of science (Bloor 1983).

This sometimes combative relation to philosophy is not the only reason the sociology of science is important. In chapter 5 I marked out three different perspectives or viewpoints on science, distinguished by how zoomed-in they are—on individuals (level 1), on scientific communities (level 2), or on whole societies in which science plays a role (level 3). Kuhn opened people's eyes to the richness of a social orientation to philosophical questions about science, but he saw scientific communities as rather self-contained and didn't have much to say about the third level. The sociology of science developed theories that explicitly cover the connections, including influences in both directions, between science and society as a whole.

7.2 Robert Merton and the "Old" Sociology of Science

Sociology is the study of human social institutions and patterns in social behavior. Science itself is surely an interesting kind of social behavior. The sociology of science developed in the middle of the twentieth century. For a while it had little interaction with philosophy of science, though that eventually changed. The founder of the field, and the central figure for many years, was Robert Merton.

Merton was very interested in the ways that particular kinds of societies fostered or hindered modern science in its early years. But I am going to look here at another side of his work, which concerned the internal workings of scientific communities. In the 1940s Merton investigated science by giving a description of its norms. It is common in sociology to describe the norms—the values and rules, perhaps implicit—that are accepted in an area of human social life. Merton said that the norms of science are *universalism, communism, disinterestedness,* and *organized skepticism.* Universalism is the idea that the personal attributes and social background of a person are irrelevant to the scientific value of the person's ideas. Communism involves the common ownership of scientific ideas and results. Anyone can make use of any scientific idea in his or her work; the French are not barred from using English results. The norm of disinterestedness is made questionable by some of Merton's own later ideas, but the idea is that scientists are supposed to act for the benefit of a common scientific enterprise rather than for personal gain. Organized skepticism is a community-wide pattern of challenging and testing ideas instead of taking them on trust. (Merton sometimes added *humility* to his list of norms, but that one is less important.)

Merton looked at scientific norms from another angle in a famous (and wonderfully readable) paper first published in 1957. This is Merton's account of the *reward* system in science.

Merton claimed that the currency of scientific reward is *recognition,* especially recognition for being the first person to come up with an idea. This, Merton claimed, is the only property right recognized in science. Once an idea is published, it becomes common scientific property, according to the norm of communism. In the best case, a scientist is rewarded by having the idea named after him, as we see in such cases as Darwinism, Planck's constant, and Boyle's law.

Merton argued that the importance of recognition is made evident by the fact that the history of science is crammed with priority disputes, often of the most acrimonious kind. And let no one suggest that the figures involved were just the jealous also-rans. Galileo fought tooth and nail over recognition for various ideas he saw as his own. Newton fought Hooke over the inverse-square law of gravity, and he fought Leibniz over the invention of calculus. Newton was so eccentric and so absorbed in his research that he had to be reminded by his servant to eat, but he did not have to be reminded to get into priority disputes.

The pattern has continued similarly from the seventeenth century to today. National loyalties are sometimes a factor, as seen in the dispute between the American Robert Gallo and the Frenchman Luc Montagnier over the discovery in 1983 of HIV, the virus responsible for AIDS. There are some exceptions to the general tendency; the most famous is the tremendously polite and gentlemanly non-dispute between Charles Darwin and Alfred Russel Wallace over the theory of evolution by natural selection in the middle of the nineteenth century. But the usual pattern when two scientists seem to hit on an idea around the same time is to fight for priority. As Merton says, the moral fervor seen in these debates, even on the part of those with no direct involvement, suggests that a basic community standard is operating.

Some priority disputes have recently taken on another dimension. In molecular biology, new discoveries quite often have potential for commercial applications, with huge amounts of money at stake. Around 2012, the method for editing genes known as CRISPR was developed (partly by putting together earlier discoveries about bacteria) by two teams, one led by Jennifer Doudna and Emmanuelle Charpentier, the other by Feng Zhang. Even before an academic priority dispute had warmed up, both sides had filed patents, set up biotechnology companies, and started a court battle over the commercial side of this work. In the parts of science that have potential for commercialization, the nature of priority disputes has changed considerably. Some of this was visible also in the dispute about the discovery of HIV between the French and American teams some decades earlier. A valuable blood test would come out of that work, and the test would make someone a lot of money. In the end, the HIV/AIDS priority dispute was resolved in large part by an official agreement not between scientists, but between the presidents of the United States and France at the time, Ronald Reagan and François Mitterand. The agreement split the credit (and the money that was to come) fifty-fifty.

It is not surprising that intense priority disputes occur when large sums of money are waiting to be given to someone. What is more surprising is that scientists have, over centuries, expended enormous amounts of energy in these disputes when there is (or seems at the time to be) no money at all at stake.

Merton argued that the reward system of science mostly functions to encourage original thinking, which is a good thing. But the machine can also misfire, especially when the desire for reward overcomes everything else in a scientist's mind. The main "deviant" behaviors that result are fraud, plagiarism, and libel and slander. Of these, Merton held that fraud is very rare, plagiarism somewhat less rare, and libel and slander very common. Fraud is rare, he said, largely because of a rigorous internal policing by scientists, which derives partly from their own ambitions but also from organized skepticism. (Fraud may have become more common since Merton's time; I'll return to this in a later chapter.) Plagiarism does happen, but the most usual outlet for deviant behavior is libel and slander of competitors. More precisely, what we find is a special form of slander that relates to the reward structure of science: *accusation* of plagiarism. This is vastly more common than actual theft. When two scientists seem to hit on an idea simultaneously, it is easy and often effective to insinuate that while *my* discovery was legitimate, it is no accident that Professor Z unveiled something very similar around the same time. After all, Professor Z took a lot of notes during an informal talk I gave some months ago, and he also cornered one of my graduate students and wanted to know how we had managed to ... (etc.). Professor Z, or Z's allies, will often reply in kind.

Merton also has a poignant discussion of the fact that the kind of recognition that is the basic reward in science will be given to only a small number of scientists. There are not enough laws and constants for everyone to get one. The result is mild forms of deviancy such as the mania to publish. For pedestrian workers who cannot hope to produce a world-shaking discovery, publication becomes a substitute for real recognition.

Though the mania to publish is certainly real (and no less so in philosophy than in science), I suggest that Merton's analysis is not quite right on this point. Scientists (and philosophers) who cannot hope to produce another $E = mc^2$ will nonetheless often have real standing in a small community of people who work on the same problems. Recognition even in a tiny community of colleagues can be a significant source of motivation. Kuhn's analysis of normal science recognized this fact. And in explaining the mania to publish, at least in recent years, university administrations and their desire for ways to measure productivity play a role. I'll come back to these issues again in chapter 9, and discuss an account of incentives in science that improves on Merton's.

When I first read about these features of scientific behavior, in Merton and others, all this seemed to me to be a rather peculiar kind of human motivation. Scientists seemed to be unusual people. It is rare for scientists to steal their grant money to buy fancy cars or even lavish lunches. That cannot be said of various other professions. Scientific misdeeds tend to be more "internal." I remember reading about this in the 1980s and having a sense that Merton's analysis seemed largely right, and that scientists were weird. They seem a lot less weird now, in the days of the Internet and social media. Now a lot of people spend a great deal of their time and money in what seems a rather pure quest for recognition, and there are new forms of Merton's "mania to publish." The scientific obsession with credit looks a lot less unusual now.

Merton's editor, Norman Storer, suggests that we think of Merton's four norms as being like a motor, and think of the reward system as the electricity that makes the motor run (Merton 1973). The norms describe a structure of social behavior, and the reward system is what motivates people to participate in these activities. But it is not so clear what the relation is between the two parts of the story. As Merton himself noted, the reward system can be in tension with the norms. In fact, I cannot see what remains of the norm of disinterestedness if we accept Merton's analysis of the reward system. What we seem to have is not disinterestedness, but a special kind of ambition and self-interest.

Earlier discussions in this book also suggest possible problems with Merton's "organized skepticism." There is something right in this idea, and it has the same kind of intuitive appeal seen in simple statements of empiricism. But we must confront Kuhn's argument that too much willingness to revise basic beliefs makes for chaos in science. What we find in science, Kuhn claimed, is a delicate balance between skepticism and trust, between open-mindedness and dogmatism.

Still, we see in Merton's analysis a good pattern for a theory of the structure of science. We have a description of the rewards and incentives that motivate individual scientists, and we have an account of how these individual behaviors generate the higher-level social features of science.

Merton's sociology is often seen as the "old" style of sociology of science, a style that was superseded about forty years ago. But I think there are some very good ideas here, especially in the treatment of rewards. And sociology of science in this tradition does continue, albeit with less drama and gnashing of teeth than we find in the newer approaches.

7.3 *The Rise of the Strong Program*

The sociology of science changed, expanded, and became more ambitious in the 1970s. This is one of the fields that Kuhn's work transformed. Among other things, the field changed its overall goals. Earlier work wanted to describe the social structure of science, especially its institutions and norms, but did not try to explain the content of science—what is believed in various scientific communities. The newer approaches have often tried to use sociological methods to also explain why scientists believe what they do, why they behave as they do, and how scientific thinking and practice change over time.

Merton, in at least some of his work, assumed a view of scientific theories that seems close to logical empiricism—theories are networks of predictive generalizations. The overall picture suggested was one where once a particular kind of community is in place—a community the sociologist describes—the development of ideas will be shaped mainly by data, in a way that fits with empiricism. Kuhn's ideas about scientific change, holism about testing, the theory-ladenness of observation, and related themes opened up more ambitious projects for sociologists looking at science. Some sociologists saw their work as replacing philosophy of science, a field that had become dried up and burdened with useless myths.

The first influential project in this new approach is called the *strong program in the sociology of scientific knowledge.* This project was developed by an interdisciplinary group based in Edinburgh, Scotland, in the 1970s, headed (to some extent) by Barry Barnes and David Bloor. A central idea of the strong program is the "symmetry principle." This principle holds that all forms of belief and behavior should be approached using the same kinds of explanations. In particular, we should not give totally different kinds of explanations for beliefs that we think are true and beliefs that we think are false. Our present assessment of an idea should have no effect on how we explain its history and social role.

Applied to science, the symmetry principle tells us that scientific beliefs are products of the same general kinds of forces as other kinds of belief. Scientists are not some special breed of pure, disinterested thinkers who receive private bulletins from Nature. People of all kinds live in communities that have socially established local norms for regulating belief—norms for supporting claims, for handling disagreement, for working out who will be listened to and who will be ignored. These norms will often be subtle habits rather than explicitly stated rules.

Scientists work in a rather unusual kind of local community. This community is characterized by high prestige, lengthy training and initiation, and expensive toys. But, sociologists insist, it is still a community in which beliefs are established and defended via local norms that are human creations, maintained by social interaction. Scientists often look down on beliefs found in other communities, but this disparaging attitude is part of the local norms of the scientific community. It is one of the rules of the game.

As a result, we must recognize that the general kinds of factors that explain why scientists came to believe that genes are made of DNA are the same kinds of factors used in explaining how other communities arrive at their very different beliefs—for example, a tribal community's belief that a drought was due to the ill will of a local deity. In both cases, the beliefs are established and maintained in the community by the deployment of local norms of argument and justification. The norms themselves vary between the tribal community and the community of scientists, but the same general principles apply in both cases. Most important, we should not give the *Real World* a special role in the explanation of scientific belief that it does not have in the explanation of other beliefs that pass local community norms.

Here is a quote from a more recent book—not one the sociologists were responding to—that is an especially clear example of what they were against. It is from the psychologist Steven Pinker, who is writing about progress in different areas of life:

> [W]e don't have to explain why molecular biologists discovered that DNA has four bases—given that they were doing their biology properly, and given that DNA really does have four bases, in the long run they could hardly have discovered anything else. (2011, 180)

The idea here is that if a community came to believe that DNA has six bases, then we'd have to give a detailed explanation of how this could happen: they got confused, their samples were contaminated—something like that. But if a community believes DNA has four bases, and that community is made up of competent scientists, there is not much that needs explaining.

This attitude, the sociologists thought, is a big mistake. We might today think that some scientific idea (DNA has four bases) is right, or justified, but that fact can't have any role in the past processes by which the idea got established. A reply to this might be as follows: yes, our present judgments can't affect the past development of an idea, but a single set of facts about the experimental support and reasonableness of an idea can affect both the past assessment of the idea and our own assessment today. It's misleading to talk about present judgments affecting earlier events; that is a caricature. But what lies behind our judgment now—how DNA actually is, and how those facts about DNA affect the experimental work we do—can also have affected what people thought back then.

That would not be the end of the argument. Maybe something about actual DNA causally affected what people thought back then, as well as now, but what people think and say now is affected by what people decided about DNA back in the middle of the twentieth century. There is not one causal path from the actual nature of DNA to people's beliefs about DNA in the 1950s, and a separate causal path from the nature of DNA to our beliefs about it now; the whole situation is more tangled up than that.

This debate could continue, but I'll get back to the development of the sociology of science. With the symmetry principle as a guide, the strong program not only sought to describe how local norms lead to some ideas being accepted and others rejected, but also tried to analyze particular scientific theories in terms of their relations to their social and political context. This work became especially controversial. The aim was to explain some scientific beliefs in terms of the political "interests" of scientists and their place within society.

For example, Donald MacKenzie (1981) argued that the development of some of the most important ideas in modern statistics should be understood in terms of the role these tools had in nineteenth-century English thinking about human evolution and its social consequences. That connection went partly via the program of *eugenics*, the attempt to influence human evolution by encouraging some people to breed and discouraging others from doing so. MacKenzie argued that a body of biological, mathematical, and social ideas was well matched to the interests of the ambitious, reformist English middle classes. He was asserting some kind of link between the popularity of specific scientific and mathematical ideas, on the one hand, and broader political factors on the other. What kind of link was this supposed to be? MacKenzie was cautious. When links are made between specific scientific ideas and their political context, the sociologist of science is often quick to say that no simple determination of scientific thinking by political factors is being alleged. Sometimes metaphorical terms like "reflect" are used: scientific ideas will "reflect the interests" of a social group. Certainly, it can sometimes be shown that the popularity of a scientific idea benefits a social group. But is this benefit supposed to explain the popularity of the scientific idea or not? If so, is the explanation supposed to be a causal explanation, albeit a qualified one, or some other kind? This has been the source of some obscurity, but the issue of causal analysis in complex social systems is often very difficult. Some kind of explanation is intended, though.

This work on scientific ideas and interests certainly antagonized conventional philosophers and historians, but it didn't only antagonize people who were the targets of these ideas. Even Kuhn was critical of it. Although Kuhn's work is always cited by those seeking to tie science to its broader political context, his book *Structure* did not have much to say about the influence of "external" political life on science. Kuhn analyzed the internal politics of science—who writes the textbooks, who determines which problems have high priority. For Kuhn, an insulation of scientific decision-making from broader political influences is a strength of science. Despite his status as a hero, Kuhn did not like the radical sociology of science that followed him.

The strong program is also often associated with *relativism*. Many sociologists accepted this label, but we need to be cautious with its use. There are so many definitions of relativism floating around that the sense of relativism embraced by the sociologists need not be the same as that used by commentators and critics. The forms of relativism that are important here concern standards of rationality, evidence, and justification. Relativism in this context holds that there is no single set of standards entitled to govern the justification of beliefs. The applicability of such standards depends, instead, on one's situation or point of view (see also the Glossary and section 5.6). In this sense, the strong program does tend to be relativist. It holds that science has no special authority that extends beyond all local norms. Instead, the norms and standards that govern scientific belief can be justified only from the inside, and that is true of other, nonscientific norms as well. We who live in science-dominated societies will find it compelling to say, "Science really is the best way of learning about the world." But, according to many sociologists of science, saying those things is just expressing our local norms. No one can hope to take a point of view outside all local norms and conceptual systems and say, with any justification, that some particular conceptual system or this set of local norms really is the best, the one that adapts us best to the world.

Despite some differences within the field, it is fair to say that the strong program is an expression of a relativist position about belief and justification. A famous problem for relativists is the application of relativism to itself. The problem does have various solutions, but it can definitely lead to tangles. Unfortunately, that is what happened in sociology of science. The application of the field's principles to itself led to interminable discussions that weighed the field down. If all beliefs are to be explained in terms of the same kinds of social factors, and no set of local norms can be judged "really" superior, from an external standpoint, then what about the theories found in sociology of science? This came to be called the "problem of reflexivity." Mostly, the sociologists of science accepted that their claims were true of their own ideas. They accepted that their own theories were only justified according to local social norms. This conclusion is OK, but the issue led to a great deal of methodological obsessing and navel-gazing.

The strong program was not the only kind of sociology of science developing in this period. Just as the strong program elbowed aside earlier social accounts of science in the 1970s, it was to be partially elbowed aside in turn, in the 1980s.

7.4 Leviathan, *Latour, and the Manufacture of Facts*

This section will look at two particularly influential books in the sociology of science and at how the field changed in the latter part of the century. The first is Steven Shapin and Simon Schaffer's *Leviathan and the Air Pump* (1985; from here on I will refer to it as *Leviathan*). This is a work of sociologically informed history, rather than pure sociology. It does not advocate the strong program, but it is often seen as a sophisticated development of those ideas. The second work is central to another shift that took place in sociology of science. This is Bruno Latour and Stephen Woolgar's *Laboratory Life* (1979).

Leviathan discusses the rise of experimental science in seventeenth-century England. This was a pivotal time in establishing the social structure of modern science. The book focuses on a dispute between Robert Boyle, a leader in the new experimental science, and Thomas Hobbes. Hobbes is remembered now mainly as a political philosopher (Hobbes's 1660 book *Leviathan*, which Shapin and Schaffer refer to in their title, is a work of that kind), but he also engaged in scientific disputes. The battle between Boyle and Hobbes was not a case of "science versus religion," or anything like that; it was a battle over some specific scientific issues and over the right form for scientific work to take. Boyle prevailed.

According to Shapin and Schaffer, what came out of this period, and especially from Boyle's work, was a new way of bringing experience to bear on theoretical investigation. Boyle and his allies developed a new picture of what should be the subject of organized investigation and dispute, and how these disputes should be settled. The Royal Society of London, founded in 1660 by Boyle's group, became the institutional embodiment of the new approach. Boyle's approach did not become the only model for science during the later seventeenth century, but it was one very important model, especially in England. There were some fairly strong differences in scientific "style" between different European countries during this period (and many would say that these have not entirely disappeared).

Boyle sought to sharply distinguish the public, cooperative investigation of experimental "matters of fact" from other kinds of work. Proposing causal hypotheses about experimental results is always speculative and should be done cautiously. Theological and metaphysical issues should be kept entirely separate from experimental work.

In marking out a specific area in which dispute could be controlled and productive, Boyle hoped to show that scientific argument was compatible with social order. The seventeenth century had seen civil war in England, and this whole period in European history was one in which even the most abstract theological questions seemed capable of leading to violent unrest. So there was much concern with the problem of how to control dissent and dispute—how to stop it from spilling over into chaos. According to Shapin and Schaffer, Boyle saw his group of experimentally minded colleagues as a model for order and conflict resolution in society at large.

Boyle was not only setting up new ways of organizing work; he was also setting up new ways of talking—new ways of asking and answering questions, handling objections, and reaching agreement. We see this, Shapin and Schaffer argue, in Boyle's handling of key terms such as "vacuum." The existence of vacuums was a frequent topic of debate in the seventeenth century. Aristotle's physics held that vacuums could not exist, but various lines of experiment suggested that perhaps they could. Boyle's experimental work involved the use of a pump that could apparently evacuate all or almost all the air from a glass container, in which experiments could then be performed. Shapin and Schaffer argue that Boyle was not trying to answer the standard, familiar questions about vacuums. Instead, he was reconstruing questions about vacuums in a way that brought them into contact with his experimental apparatus. Critics could—and did—complain that Boyle's pump could not settle the questions they wanted to ask. Boyle's strategy was to subtly replace these questions with other questions that could be the topic of experimental work. The old questions—such as whether an absolutely pure vacuum could exist—had been set up in such a way that they would generate endless and uncontrollable dispute.

Shapin and Schaffer present their view in terms taken from the (later) philosophy of Ludwig Wittgenstein. Since Wittgenstein has influenced many people in the sociology of science, it is worth taking a moment to sketch some of his ideas.

Wittgenstein's early work developed a formal model of how language works and how sentences relate in a picture-like way to "facts." He invented some tools still used today in formal logic, and his work strongly influenced logical positivism. His later ideas, especially his *Philosophical Investigations* (1953), were very different, and they had huge effects on late twentieth-century thought. These later ideas are more an "anti-theory" than a theory; they are an attempt to show that philosophical problems arise from pathologies of language. Philosophy arises from a subtle transition between ordinary use of language and a kind of linguistic misfiring, in which questions that are really incoherent can seem to make sense. Wittgenstein wanted to diagnose and put an end to these misguided linguistic excursions. He avoided formulating definite theories of anything, but some of his ideas have been adapted for use in theories in various areas, including sociology of science (Bloor 1983).

Two ideas are especially popular. A "form of life," for Wittgenstein, is something like a set of basic practices, behaviors, and values. Actions and decisions can make sense within a form of life, but a form of life as a whole cannot be justified externally. It's just the way a group of people live. Wittgenstein was not much interested in the kinds of cultural variation studied by sociologists and anthropologists, and it is not clear what sort of unit a form of life is for him. But sociologists have adapted the concept to fit the kinds of groups they study.

The second concept drawn from Wittgenstein is the concept of a "language game." A language game is a collection of habits in the use of language that contribute to a form of life and make sense within it. Wittgenstein opposed a view of language in which words and sentences are attached to their own particular meanings (mental images, perhaps) that determine how language is used. Instead, Wittgenstein claimed we should think of the socially maintained patterns of language use as *all there is* to the "meaning" of language. Shapin and Schaffer argue that Boyle's treatment of crucial terms like "vacuum" established a new language game. This language game was a key component in a new form of life, the form of life of experimental science.

At this point you may be remembering the logical positivists and their attempt to analyze the meaning of scientific language in terms of patterns in experience. Is the idea of a language game, developed to serve experimental science, different from this positivist idea? It is different. The logical positivists claimed that the right theory of meaning would show that the only thing meaningful language ever does is describe patterns in experience. According to Shapin and Schaffer, in contrast, Boyle was setting up a new way of using language. So there is perhaps more of a connection to the "operationalism" of the physicist Bridgman, who was briefly mentioned in chapter 2. Bridgman (1927) urged that scientists reform their use of language to ensure that each term has a direct connection to empirical testing.

There's another side of the view presented in *Leviathan* that is more philosophically provocative. Shapin and Schaffer claim that Boyle and other scientists are engaged in the *manufacture of facts*. In everyday talk, the phrase "manufacture of facts" would be taken to indicate deception, but that is not what Shapin and Schaffer have in mind. For them, there is nothing bad about the manufacture of facts; they want us to get used to the idea that facts in general are made rather than found. This is reminiscent of Kuhn's claim, in chapter X of *Structure*, that the world changes during a scientific revolution. Like many others who use these terms, Shapin and Schaffer want to reject a picture of the scientist as a passive receiver of information from the world. But denying passivity does not require this kind of talk, and it often leads to trouble. For example, at the very end of *Leviathan*, their discussion of "making" leads Shapin and Schaffer to express their overall conclusions in a way that involves a real confusion. They say: "It is ourselves and not reality that is responsible for what we know" (1985, 344). This is a classic example of a false dichotomy. Neither we alone nor reality alone is responsible for human knowledge. The rough answer is that both are responsible for it; knowledge involves an interaction between the two. Even this formulation is imperfect; human knowledge is part of reality, not something separate from or outside it. But, speaking roughly, in order to understand knowledge, we need both a theory of human thought, language, and social interaction, and also a theory of how these human capacities relate to the world outside us.

I now move on to another famous work in the sociology of science, Latour and Woolgar's *Laboratory Life* (1979). In the mid-1970s, Bruno Latour, a French philosopher/anthropologist, spent a couple of years visiting a molecular biology laboratory, the Salk Institute in San Diego. He went as a charming observer who knew little about molecular biology. During the time that Latour was there, the lab did work that resulted in a Nobel Prize; they discovered the chemical structure of a hormone involved in the regulation of human growth. Latour wrote *Laboratory Life* with Steven Woolgar as a description of the lab's work.

Latour and Woolgar, in their account, ignored most of what a normal description of a piece of science would focus on. They ignored the state of our knowledge of hormones; they ignored the ways in which

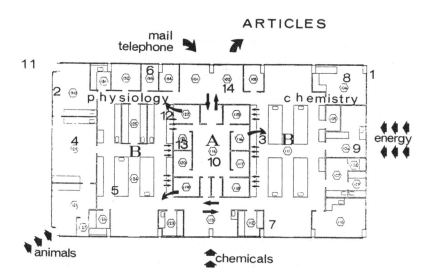

Figure 7.1. The lab according to Latour and Woolgar (From *Laboratory Life: The Construction of Scientific Facts,* by Bruno Latour and Steve Woolgar, © 1986; reprinted by permission of Princeton University Press)

experimental methods in the field are able to discriminate alternative chemical structures; they ignored how the new discovery fit into the rest of biology. Instead, Latour saw himself as an anthropologist visiting an unfamiliar tribe and trying to work out what meaning their strange behaviors had. Latour and Woolgar described the lab as a complicated device where chemicals, small animals, and reams of blank paper went in at one end, and small printed pieces of paper—journal articles and technical reports—came out at the other end. In between the two, a huge amount of "processing" went on, processing that turned the mass of raw materials into the intricate finished products (see fig. 7.1). Latour saw this processing as aimed at taking scientific claims and building structures of "support" around them, so they would eventually be taken as facts. A key step in this process is hiding the human work involved in turning something into a fact; to turn something into a fact is to make it look as though it is not a human product but is given directly by nature.

To many, *Laboratory Life* seemed like a breath of fresh air, a book that exuded wit and imagination. Along with other works, it prompted a shift within the field. The strong program came to seem crude. The strong program wanted to get rid of explanations of scientific belief in which nature just stamps itself on the minds of the scientific community. But perhaps the strong program was replacing this with an equally crude picture, in which social and political "interests" stamp themselves on the scientific community. This is not a very fair reading of the strong program; the sociologists were being caricatured, just as they had caricatured traditional philosophy of science. Some would see justice there. Latour also inspired a very different style in sociology of science, a style that might be described as elusive, self-conscious, and literary.

In Latour's approach, the sociologist studies the details of the internal dynamics of scientific work, especially disputes and negotiation about what has been established. The sociologist does not begin the story by taking for granted "pressures" or "interests" in society at large, and "nature" or the real world is not taken for granted either. Instead, both society and nature are seen as products, not causes, of the settlement of scientific controversies (Latour 1987). Where traditional empiricist philosophy had seen science as "data-driven," and the strong program had seen science as "interest-driven," Latour sees scientific work itself as the driver.

In Latour's view, when we explain why one side succeeded and another failed in a scientific controversy, we should never give the explanation in terms of nature itself. Both sides will be claiming that they are the ones in tune with the facts. But when one side wins, that side's version of "the facts" becomes *immune to challenge*. Latour describes this final step as a process in which facts are created, or constructed, by scientific work. Scientists, through their decisions, build networks of support around some claims, making it harder to question them and easier to accept them. This work includes hiding the human choices involved; to turn something into a fact is to make it look like it was just "given" by nature.

There is an interesting strategy here, along with much obscurity. Latour gets us to look at the dynamics of controversy in science in a very fine-grained way. What social role is played by appeals to "truth," "nature," and "the facts"? How do people use these terms before, during, and after the settlement of disputes? These are good questions. One kind of understanding we might have of the concepts of "truth" and "nature" is an understanding of how they are used as resources in arguments and discussion. But while an investigation of this kind might tell us a lot about how people decide what they *take* to be real, this does not mean that the settlement of scientific disputes determines what *is* real. Latour's books are full of artful avoidance of this distinction.

From around 2016 or so, Latour himself began presenting a different message about science. Concerned about climate change, and the problem of public doubts about it, he began to express more explicit admiration for science and confidence in its products. To many, Latour seemed to be rejecting his earlier work, which encouraged a less trusting attitude toward science. Latour has responded that he is not taking back his older ideas. Those were ideas about how scientific belief is formed. He wanted to get inside the workings of science. Having a detailed view of those workings need not mean that you are opposed to what comes out. However, we might reply to Latour that some views of the workings of a mechanism do encourage a distrust in what comes out, and Latour's analyses did indeed have that form, before his change of mind.

Setting aside Latour himself, the idea of manufacture or construction of facts is a feature of many radical views about science. What is going on here? I think at least two things are going on (and I'll come back to this again in chapter 10, the official discussion of reality in this book). One thing going on is a keen interest in how talk of "truth" and "reality" actually works in debates. Sociologists and others rightly think there is a lot to say about concepts like these, and how people use them in attempts to persuade others and mark out the terrain of reasonableness. But this interest in a good problem is often combined with an unwillingness—sometimes via a taste for dramatic pronouncement—to distinguish between claims made about our view of the world and claims made about the world itself.

Another thing operating in discussions of the manufacture or construction of facts is a more radical motivation. Some writers want to explore a kind of deliberate *inversion* of the usual kind of story that might be told about our beliefs and the objects those beliefs are about. Rather than there being a path from objects through perception to beliefs and the things we write down and say to each other, perhaps the path runs the other way; perhaps the objects that make up the world are a consequence of what we think and say.

Why might you try to invert the relationship? Is it just a matter of wanting to make a splash? I think the motivation for many has been a strong sense that the more familiar picture has actually collapsed. There was an old way of seeing the overall situation—with objects affecting our minds through perception, leading to beliefs about the objects—and for many that picture has collapsed. (Kuhn called the old view an "epistemological paradigm.") As a result, theorists were looking to build something completely different, and that has led to an exploration of all sorts of inversions and reversals of the usual relationships. (Often these reversed pictures made use of the same ingredients as before—was this a good idea, if the goal was something really new?)

Suppose we want to do things differently—not assert a simplistic view in which facts stamp themselves on scientific minds, but not assert a complete reversal of that picture either. How might this go? Let's work a little from Latour and Woolgar's own example in *Laboratory Life*. We have a lab in San Diego, with tiny traces of chemicals (hormones) inside mice. Latour and Woolgar's view is built around the idea that scientists are not passive recipients of information about these chemicals; scientific life is full of wrangling, negotiation, and trade-offs, full of decisions about who to trust and how to persuade. These, Latour and Woolgar insist, are all *choices*. Yes, they are, but one thing scientists can choose to do is to have what they say guided by how experiments, measurements, and assays turn out. They can decide: if I see X, I will think and do this; if I see Y, I will think and do that. They might make this choice and then later change their mind or fudge things. But they also might not. If they do make decisions of this kind and stick to them, the world inside the mice then plays some role in constraining what scientists say. Whether we learn much about the world outside the San Diego mice—for example, other animals—is a further question. Questions about extrapolation and induction will not stay away for long. But this is a start.

Further Reading and Notes

A standard collection of Merton's works is *The Sociology of Science* (1973). Website [2] updates the HIV priority dispute and its aftermath. I'll come back to credit and reward in science in chapters 9 and 14.

A central work in the rise of the strong program is Bloor's *Knowledge and Social Imagery* (1976). See also Barnes, Bloor, and Henry (1996). Shapin (1982) is a good survey of historical work done by sociologists of science. On the issue of relativism, see Barnes and Bloor (1982) and other papers in that collection (Hollis and Lukes 1982). Latour's other famous books include *Science in Action* (1987), *The Pasteurization of France* (1988), and *We Have Never Been Modern* (1993). The apparent shift around 2016 in Latour's view of science has been discussed in several interviews with him. One is website [3].

For work on the social role of appeals to truth and reality, see Shapin's *A Social History of Truth* (1994). Daston and Galison take a historical look at the concept of *objectivity* in their 2007 book of that name.

A note about the last paragraphs of this chapter: the word "fact" causes some trouble here. "Fact" can be used to mean either *an arrangement in the world* or *a sentence that is true*. These two meanings are quite different. Any sentence is manufactured or constructed, but its truth, and the things it is about, often are not. All this will be discussed further in chapter 10.

Chapter 8

Science Is Political

8.1 *A Changing Image of Science*

The relationship between science and politics was subjected to new kinds of scrutiny in the late twentieth century. To some extent, the overall image of science changed, especially within the humanities. Generalizations are risky here, but we might say that through much of the last three hundred years, science has been regarded in Western societies as a progressive, anti-authoritarian force, able to challenge and break down entrenched ideas and arrangements. This view was most vivid in the Enlightenment period, in the eighteenth century, and confidence in science as a progressive force is part of a collection of ideas now known as "Enlightenment values." There have always been exceptions to this image of science—within some Marxist thought, for example—but many parts of intellectual culture saw a larger shift in attitudes toward science in the latter part of the twentieth century.

Science came to be seen instead as a force in the maintenance of the status quo, especially with respect to political inequalities. On the side of politics that considers itself progressive rather than conservative, many began to treat science as part of a larger, multi-tentacled political structure that acts to reinforce subtle forms of exclusion and coercion, even in apparently free and democratic societies. The anti-authoritarian image of science came to be seen as just "good PR." And the institution of science itself, it was argued, is full of hidden features that exclude some individuals while welcoming others. Revealing the connections between scientific institutions and political power would show that "science is political," rather than being an institution outside of politics that enjoys a special authority derived from neutrality. Making this clear would also have relevance to questions about education, medicine, and a variety of other aspects of social policy.

The most important manifestation of this new attitude is found in some forms of feminist philosophy of science. That is the topic of the first part of this chapter. The second part looks at the development of an interdisciplinary field that puts a number of these critical perspectives on science together. The last part of the chapter looks at broader questions about science, values, and practical projects.

8.2 *The Man of Reason*

Feminist thinking about science is diverse, but it has been unified to some extent by the idea that science has been part of a structure that has perpetuated inequalities between men and women. Science itself, as well as mainstream theorizing about science and knowledge, has helped to keep women in a second-class position as thinkers, investigators, and intellectual citizens. (Even these generalizations about feminist discussions of science have exceptions.) According to feminist analyses, society has suffered from this, and so has science itself. So reform of some kind is needed. There is disagreement on the appropriate kind of reform—ranging from simple suggestions like the recruitment of more women in the sciences, through changing the scientific career path—which does tend to favor men—through to the encouragement of a specific kind of female "voice" in science, and, most radically, to dethroning science from its preeminent position in Western culture.

We should distinguish feminist *philosophical* ideas about science from broader feminist *political* ideas. Feminism in general aims to understand and fight against inequalities between the sexes in political rights, economic standing, and social status. This has a simple application to science: women were for many years excluded or discouraged from a life in science, as they were excluded from other high-prestige areas of work. So far, this might be seen as a familiar but important matter of equality of opportunity, one that raises questions about policy (such as the appropriateness of affirmative action) but does not raise issues in the philosophy of science itself.

Other feminist work did engage with philosophical issues about science. The work might be categorized in terms of three overlapping projects. The first is feminist analysis in the history of ideas and the history of science. The second is feminist analysis of specific scientific fields and theories, especially in social science, biology, and medicine. The third is *feminist epistemology*, the attempt to analyze rationality, knowledge, and other basic epistemological concepts from a feminist point of view.

I will start by discussing a book written fairly early in the tradition,

Genevieve Lloyd's *The Man of Reason* (1984). Lloyd analyzes the historical roots of ideas about knowledge and rationality—in figures such as Plato, Aristotle, Descartes, and Bacon—and also draws conclusions for epistemology. The book illustrates what I think has been a common pattern. Lloyd tells a very interesting story in the history of ideas, one that is sometimes quite compelling. It is harder to work out what consequences these historical facts have for philosophy.

Lloyd argues that the historical development of ideas about reason and knowledge was greatly affected by views about the relation between maleness and femaleness. The concept of reason evolved in Western philosophy in a way that associated reasonableness with maleness, and associated the female mind with a set of psychological traits that *contrast* with reasonableness. An important source for this pattern of thinking, according to Lloyd, is the old association between femaleness and nature; the Earth is fertile, female, the source of life. Via this association, ideas about the relationship between the mind and nature were modeled on the relationship between male and female. The relations between the sexes also provided a model for theorizing about the relations between different aspects of the mind itself—between perception and thought, and between reason and emotion. The upshot was that the ideas feeding into the early development of science and philosophy in Europe incorporated, in various different forms, an association between reason and maleness. And the development of the idea of femininity was shaped by an opposition between femininity and reason. Femininity was associated with receptivity, intuition, empathy, and emotion.

Lloyd's best example is Francis Bacon, the seventeenth-century English thinker who wrote extensively about the new empirical methods of investigation and their promise. I discussed some of his ideas about scientific methods near the end of chapter 3. Bacon attacked the ancient Greek picture of knowledge as contemplation. For Bacon, real knowledge is manifested in control of nature: knowledge is power. But as Bacon developed this idea, he retained the image of nature as female. His model for the relation between the mind and nature was the model of marriage, a marriage between the knower (man) and nature (woman). The features of a good marriage, as run by the man, correspond to the features of successful knowledge of the world. So what is a good husband like? A good husband is respectful, but he is also firm and definitely in charge. The scientist approaching nature should approach her with respect and restraint, but control is certainly needed: "Nature betrays her secrets more fully when in the grip and under the pressure of art than when in enjoyment of her natural liberty." The products of what occurs there on the "nuptual [*sic*] couch" will be useful knowledge for the improvement of mankind (quoted in Lloyd 1984, 11–12). Back in chapter 2, using a quote from the physicist Richard Feynman, I looked at the distinction between passive observation and active experimental work. Bacon, unlike many other empiricist philosophers, emphasized the importance of this distinction. He also described it in sexualized terms.

Cases like this suggest that views about the relations between men and women were important resources in the development of ideas about reason and knowledge. Although questions of causation in these large-scale historical matters are always difficult, it surely seems likely that these associations did affect both the lives of women and the path taken by science in the early modern period. The harder question is what philosophical consequences these historical facts have for us now, given the massive changes to political life and to science since then. It is not hard to find a residue of these old associations embedded in metaphors that are still around today. To pick a simple case, scientists sometimes talk about whether or not a phenomenon will "yield" to a particular method of analysis. To my ear (though not to everyone's), this metaphor always has a resonance of sexual conquest. But whether these metaphors have much effect on either society or science today is a more difficult issue.

Evelyn Fox Keller, another prominent feminist, thinks there is a real problem here. She holds that as a result of the image of science that has been established, the woman scientist has to choose between "inauthenticity" and "subversion." This concept of authenticity is a subtle one drawn from existentialist philosophy, but Keller illustrates her point with an analogy: "Just as surely as inauthenticity is the cost a woman suffers by joining men in misogynist jokes, so it is, equally, the cost suffered by a woman who identifies with an image of the scientist modeled on the patriarchal husband" (2002, 134–35).

8.3 *Sex and Gender in Behavioral Biology*

In this section I will look at some particular parts of science where issues of gender have been vexed and controversial. My main case will be one that many see as a clear example of how the gender of researchers has had an effect on the development of ideas, one where science has benefited from an increasing role for women in the field. The example concerns the study of social behavior, especially sexual behavior, in nonhuman primates like chimps and baboons. These phenomena are studied (with slightly different emphases) in the fields of primatology and behavioral ecology.

These parts of biology initially developed a picture of primate sexual life in which females were seen as rather passive. Social and sexual life were regarded as controlled, sometimes cruelly, by males. That picture was linked to some influential pieces of "high theory" in evolutionary biology. In many animals, although by no means all, there is more variation across individuals in male reproductive success than there is in females. This is a consequence of the fact that one male can, in principle, impregnate large numbers of females. Female reproductive success is limited in many animals by the high costs of pregnancy.

This asymmetry between the sexes is of considerable evolutionary importance in the organisms in which it is found. But it has often been used in rather simplistic patterns of explanation, without regard for many ways in which its effects can be modified by other factors. In early primatology, it was taken to support a view holding that male sexual behavior had been finely honed by natural selection, while female behavior had not, because females could do much less to affect their reproductive success.

According to Sarah Blaffer Hrdy (2002), this picture began to shift in the 1970s. Careful observation revealed a more active and complex role for female primates. It became apparent that many female primates have elaborate sex lives, involving a lot more different kinds of sexual contact than one would expect based on the old picture. Females seem to engage in subtle patterns of manipulation of male behavior. The basic theoretical idea that the high potential variance in male mating success has effects on the evolution of behavior still stands, but there is now a more sophisticated picture of the interaction between this factor and others, especially the strategies available to females.

This shift in thinking within primatology coincided, at least roughly, with an influx of women into the field. Primatology is one of the scientific fields in which the presence of women is unusually strong. What role did the presence of women have in changing opinions within the field? According to Hrdy (and according to others I have spoken to), the idea that this increasing representation of women had a significant role in shifting people's views about primate behavior is widely accepted within primatology. Hrdy adds that this view seems to have had more acceptance in the United States than in Britain (2002, 187). And though Hrdy herself is cautious about this issue, she suggests that women researchers, such as herself, did tend to empathize with female primates and watched the details of their behavior more closely than their male colleagues had.

Now that sex roles in animals are on the table, I'll follow up a little, though this next comment involves different themes than women's participation in science. The established picture I sketched above—one with high variation in reproduction among males and lower variation in females—has encountered a number of surprises in recent years. I think that most biologists would still say it is an important difference between the situations of males and females in many species, but the way it has often led to general pictures of maleness and femaleness in sex roles has been criticized. Some of the biggest surprises have come from birds, especially because in some cases they initially *seemed* to follow the rules but did not. A "lek" is a pattern of social behavior (not always in birds—many species have leks) where males gather to display to females, females come and mate with some of them, and the males provide no parental care at all—no contribution to reproduction other than sperm. At leks, there seems to be a huge variation in mating success across males. But it turns out that this is not always what's going on. We might observe a few males mating much more than others, but we can also now do paternity tests on the eggs or baby birds that result and see how many are really the offspring of those apparently dominating males. It has turned out in some cases that many more males have offspring than had been suspected, and variance in success between males is quite low (Lanctot et al. 1997 is one example; see also Fine 2016). There was a lot of hidden mating going on behind the scenes. I don't know of any claim that changes in our views about sex in birds have been due specifically to contributions of female researchers, in the way that many people think applies in primates, though there are certainly a lot of women among bird biologists. The introduction of inexpensive genetic testing seems to have been crucial in this case. But it is another area in biology where an initially simple view of sex roles has been changing.

8.4 *Feminist Epistemology*

Let us now look at feminist epistemology and its encounter with science. As I noted above, this tradition is diverse. It includes work that uses feminist theory as a basis for criticizing how science handles evidence and assesses theories. It also includes feminist criticism of the social structure and organization of science, where that structure affects epistemological issues. Most ambitiously, some have argued that our familiar concepts of "reason" and "truth" themselves are covertly sexist. Feminist epistemology often also goes beyond criticism to make suggestions about reform—how to make science better at finding out about the world (if that goal is to be retained), and also how to make science more socially responsible.

In discussing the options here, I will modify some categorizations used by Sandra Harding (1986 and 1996). Harding distinguishes between three kinds of feminist criticism of science. The earliest and least controversial she calls *spontaneous feminist empiricism*. This is the project of using a feminist point of view to criticize biases and other problems in scientific work, but in a way that does not challenge the traditional ideals, methods, and norms of science.

The second category is *feminist empiricism*. Helen Longino's work

(1990) is probably the most influential within this camp, and I will discuss it below. Here the aim is to revise and improve traditional ideas about science and knowledge, but to do so in a way that remains faithful to basic empiricist ideals.

The third category I will call *radical feminist epistemology*. Two main approaches might be distinguished within this group. One is what Harding calls *feminist postmodernism*. This work tends to embrace relativism: members of different genders, different ethnic groups, and different socioeconomic classes see the world in fundamentally different ways, and the idea of a single "true" description of the world that transcends these different perspectives is a harmful illusion.

The second radical approach is *standpoint epistemology*. This is not a relativist view; in a sense, it is more ambitious than that. Standpoint epistemology stresses the role of the "situatedness" of an investigator or knower—their physical nature, location, and status in the world. The idea is that while traditional epistemology has seen situatedness as a potential problem for an investigator, in fact it can be a strength. Standpoint theory holds that there are some facts that will be visible only from a special point of view, the point of view of people who have been oppressed or marginalized by society. Those at the margins will be able to criticize the basics—both in scientific fields and in political discussion—in ways that others cannot. Science will benefit from taking more seriously the ideas developed by people with this point of view. This is usually not a relativist position, because the marginalized are seen as having *better* access to crucial facts than other people have.

One of the main debates in feminist epistemology has been between forms of philosophical feminist empiricism and views that are more radical. I will start with a look at feminist empiricism. Longino, as I said above, has been very influential within this approach, especially through her book *Science as Social Knowledge* (1990). A concept she looks closely at is *objectivity*. This, for Longino, is not a false goal or something we have completely misconceived, but it does need a new analysis.

Objectivity has been traditionally understood in terms of two different features. First, beliefs or investigations are said to be objective when they have the right kind of contact with real things, with the *objects* of the belief or investigation. Second, objectivity is supposed to involve an absence of bias, and other influences that can be called "subjective." Traditionally, the first feature is achieved by means of the second. We avoid bias and distortion in our thinking and perceiving, and thereby come into the right kind of contact with the objects we are trying to study. In this traditional view, avoiding bias and distortion is not a very difficult or complicated thing, at least in principle. We just need to attend to observational data and use good rules of reasoning.

After Kuhn and other philosophical disruptions from around his time, both the features above came to seem much more problematic. We might then give up on objectivity as a goal, or rethink and rebuild it. Longino's project is to give a new and more detailed account of the second feature above—the feature I described in terms of avoiding bias, and so on. What we need to do in a new analysis is give a central role to social structure. The aim is not to describe objective individuals, but a certain kind of community, one where disagreement and evidence are handled in the right way.

Epistemology then becomes a field that tries to distinguish good community-level procedures from bad ones. Good communities include diverse points of view, ensure minority voices are heard, and resolve disputes without coercion. This is not a simple matter, especially when a consensus has to be reached. It's always possible to continue to question things and be completely even-handed even with opinions that only a few people hold, but a community with practical decisions to make usually cannot afford that luxury. (Do we require children to be vaccinated or not?) Communities have to achieve some difficult balances in areas like this. The general point is that perhaps a theory of community-level processes of this kind is a theory of what objectivity *is*, and how to get it. The story is then not told in terms of the individual thinker and perceiver—it is not a level 1 story, in the sense of the distinction I introduced earlier, but at least a level 2 story. Objectivity exists in a community. However, even if this is right, it seems that we should also say something about individual-level features. Individual behaviors—a willingness to listen, for example—are essential to getting a community of the right kind.

Are the ideas above really specific to a feminist viewpoint? They can seem much broader. Elisabeth Lloyd (1997) argues that ideas such these are applications of the best themes seen in traditional liberal thinking, especially the ideas seen in Mill's *On Liberty* ([1859] 1978). Diversity, for Mill, provides the raw materials for social and intellectual progress, via a vigorous "marketplace of ideas." But it is no accident that feminists pushed these ideas in the 1990s. The need for a more detailed treatment of these topics was conspicuous to feminist philosophers.

So far I have been following up the second of those two threads within a traditional concept of objectivity—absence of bias and distortion—rather than the first, which involves contact with real objects. Should we also try to say something new about that other aspect of objectivity? For some writers, this becomes a non-issue—a topic we should leave aside. We just give an account of how good communities handle debates. There is a world we all live in, of course, and we might hope to do better in dealing with that world if we handle disputes in the right ways, but perhaps no more than that needs to be said.

That is one possible attitude, and one that seems to me fairly common in feminist philosophy of science. Perhaps the whole idea of working out a philosophical theory of how scientific activity relates to nature or reality is an unwanted remnant of older views. But perhaps not—perhaps we might try to come up with some new ideas in that area.

Let's also look at some more radical views in feminist philosophy of science. I distinguished *standpoint theory* and *feminist postmodernism*. Their boundaries blur, and is not always clear who is in which camp. Standpoint theory, as I introduced it via Harding, holds that the special vantage point occupied by women can lead to better work being done; the experiences of marginalized people have special value. If that is right, what sort of value is this? As Longino argues, it is not likely to be a general superiority of a kind that would justify our treating a marginalized point of view as *the* most important or reliable. If some facts are more visible to the marginalized and oppressed, other facts will surely be more visible to the privileged. The experiences of the marginalized are more likely to be valuable as a special kind of input into discussion and argument. So the right way to think here is in terms of a pool of different ideas, contributed by those with different points of view. If that is right, it brings us back to Longino's picture.

Perhaps, though, we could develop a different response to a recognition of the diversity of viewpoints. We might abandon the idea that one standpoint is really better than another, even on specific issues. We could move toward a more relativist position. This is another radical option, *feminist postmodernism*. We might, in Kathleen Okruhlik's description of this view, consider "giving up altogether the endeavor to become more and more objective" and instead accept "the existence of an irreducible plurality of alternative narratives about the way the world is" (1994, 32; here Okruhlik is summarizing the view, not endorsing it).

Even if it is tempting in some contexts, a problem with this attitude is that projects of reform can be undermined. And for many, this seems a significant loss. A commitment to reform seems to require the view that some approaches and arrangements are worse than others and could be deliberately improved. This perception is leading to a tendency, as far as I can tell, for people in this field to move back toward the moderate options. There is an unwillingness to abandon the project of trying to make things better, and in a way that retains a pretty straightforward and unqualified view of what "better" means.

In a lot of the work I've discussed here, the aim is to take a distinctively feminist point of view, considering women as a relevant unit. It was always accepted that women are not in any sense a homogeneous group of people, given the roles of economic class, race, sexual orientation, and so on. A woman who is, for example, of Central American origin is not just someone who has a female viewpoint plus a Central American one. The combination will yield its own particular perspective. This is often referred to as the phenomenon of *intersectionality*; any individual lies in the intersection of a great many overlapping sets, and their experience cannot be predicted by looking just at the general features associated with each of the groups they fall into. This point is widely accepted, but many feminists do expect there to be some fairly definite patterning in the great soup of intellectual diversity that is due to gender differences.

Certainly the physical experience of being a woman or a man will make a difference to how many aspects of life are experienced. And at least for the near future, the early education and socialization of girls and boys will have this effect as well. But we might be careful of claims that go far beyond this. It is a far harder question whether the experience and viewpoint of women are systematically different from those of men in ways that are likely to matter to scientific disputes. There is a risk of lapsing into simplistic generalizations here

Some have argued that women scientists do have a distinctive way of thinking and of interacting with the world. An example is Evelyn Fox Keller's work on Barbara McClintock, the geneticist who discovered "jumping genes" that move around within the genome of an organism. The jumping-genes idea was for some time considered to be a very strange hypothesis, but McClintock turned out to be right. McClintock was very much an outsider in genetics, and Keller also argues that McClintock had a "feeling for the organism" that enabled her to do a different style of science from that of her male colleagues (1983).

Keller does not want to argue that there will be a "sharp differentiation" between women's and men's work in science (2002, 134), but she does seem to think there will be some systematic differences. Many would object to the suggestion that a "feeling for the organism" is likely to be an example, however. A case could be made that this trait is found in a great many biologists and has nothing to do with gender. Feminists themselves (including Keller) are also wary about the possibility of contributing to a stereotyping of female contributions to scientific thinking. ("We must have a woman on this team, Jim, so someone will pick up on the holistic, interconnected stuff that might be going on in these reactions!")

As well as differences in theoretical style, another possibility is that women will tend to bring a distinct kind of social interaction to scientific communities. Feminists have sometimes suggested that women are, on the whole, less competitive and more cooperative than men, though other feminists have been very wary of generalizations of this kind (Miner and Longino 1987). If there really were differences of this kind, they could be expected to have quite significant consequences for scientific communities. As discussed in the next chapter (and as we saw already via Kuhn and Merton), the relations between competition and cooperation are very important in science.

8.5 *Postmodernism and the Science Wars*

One of the main themes in this chapter and the previous one has been an expansion of the range of fields seeking to contribute to a general understanding of science. The two examples I have discussed in detail are sociology of science (chapter 7) and feminist criticism (this chapter). As well as this expansion, there has been a blurring of disciplinary boundaries. During the 1980s a number of people decided to embrace this trend and create a new approach to studying science that would draw on many different fields without worrying about whose questions were whose. The resulting field became known as "science studies." The mixture has come to include not only history, sociology, and philosophy, but also cultural anthropology, classics, economics, some parts of literary theory, feminist theory, and cultural studies. The aim has been to draw on pretty much any field that can contribute to our understanding of how science developed, how it works, and what role it has.

Philosophers, historians, sociologists, and literary theorists do look at the world somewhat differently, so we find a mixture of styles of work within science studies. But one strand of this work gave rise to a controversy that had substantial and ongoing effects on the academic world, and I'll go through how that unfolded.

Much of the most controversial work in science studies has been allied to the movement in the humanities known as *postmodernism*. I've already used the term "postmodern" in this chapter, so we are overdue for some explanation of it. Postmodernism is a family of ideas found in fields ranging from architecture through film, history, and philosophy of language (Harvey 1989). In making sense of it, a good place to start is with what it is supposed to supersede: what is *modernism*? This, too, is a family of ideas, and early in this book we spent time with some particularly good examples of modernist thinkers, the logical positivists. Modernism was as much an approach to art and design as a philosophical approach, though. Modernists wanted to shed unwanted clutter (from buildings, from intellectual discussion) and reconstruct things from simple, basic elements. They had ideals of transparency and utility. In architecture, Le Corbusier and Mies van der Rohe were central to the movement.

Postmodernism rejected all this. The movement perhaps began, at least under that name, in an approach to architecture that sought to value quirkiness and eclecticism, rejecting austere and strict principles. (Frank Gehry's work is an example of this style, and his buildings can be seen in many cities round the world.) That work began as early as the 1960s. A broader movement that united art, architecture and academic fields like philosophy was visible from the end of the 1970s, with Jean-François Lyotard's *The Postmodern Condition* ([1979] 1984). This movement included a rejection of overarching theories and "metanarratives" of all kinds, and the replacement of those arrogant presumptions by a mosaic of local theories, languages, and projects. Postmodernism allied itself with traditions that oppose the idea that language should be analyzed as a system used to represent, or "stand for," objects and situations in the world. They described and welcomed a "dissolving" not just of traditional ideas of objectivity and truth, but of the idea that stable meanings are a feature of language at all.

Sometimes postmodernists seem to be arguing that we live, right now, in a special time in history, a time when an older representational role for symbols is being replaced by a new role. The sea of images and texts in which we live, and their roles in politics and in consumer culture, has undermined ordinary representational relations between symbols and objects. In understanding the role of symbols in our lives, it is no longer useful to apply concepts like accuracy, reference, and truth: behind every symbol lies not a real object, but another symbol. At other times postmodernism has become a tremendously obscure way of arguing for extreme forms of relativism, sometimes for a kind of skepticism and do-nothingism, and for extravagant and paradoxical views about how language and reality are related. When I was an undergraduate student in the 1980s, the conflict between movements of this kind and more established intellectual styles was acute.

Science studies was rather welcoming to postmodernism and other adventurous ideas from the humanities. As I said above, the field was diverse and included a lot of work done in a different style. But the friendly relations between this new approach to science and obscure trends in the humanities affected the image that science studies came to have. And in time, there came a backlash.

The backlash occurred in the form of an attack both on science studies and the humanities more generally—science studies was seen as an emblem of a broader decline. Some of the backlash arose within science itself; scientists were alarmed at the picture of science being presented to the broader culture. The perception was that science itself was under threat. The resulting clash became known as the "science wars." Science studies, and other work covered in this chapter and the previous one, became a battleground.

Some of the attacks on this work came from the side of conservatism in political and social thought. Advocates of "traditional" education, both in K–12 schools and in universities, worried that transmission of the treasures and values of Western civilization was being undermined by radical leftist faculty members in universities and soft-minded administrators in schools. The humanities had gone to hell, and now they were trying to wreck science as well, via endless relativist bleating that "science is just another approach to knowledge with no special status."

Although some of these battles had a recognizable left-versus-right political alignment, the most influential episode did not. In 1994 an American physicist, Alan Sokal, submitted a paper to a literary-political journal called *Social Text*, which was doing a special issue on science. Sokal's paper was a parody of radical work in science studies; it used the jargon of postmodernism to discuss progressive political possibilities implicit in recent mathematical physics. The title gives a sense of the style: "Transgressing the Boundaries: Toward a Transformative Hermeneutics of Quantum Gravity." The argument of the paper was completely ridiculous and often quite funny, but Sokal's aim was serious: he wanted to see if the journal would accept and publish it. Sokal believed that this would show the field had lost all intellectual standards and would print anything that used the right buzzwords and expressed the appropriate political sentiments.

Social Text published the paper (Sokal 1996b), and Sokal revealed his hoax in the journal *Lingua Franca* (1996a), an irreverent chronicle of academic life. The uproar reverberated across the academic world and also made the newspapers. One of the things that made Sokal's attack so effective was that he was not writing from the point of view of conservative politics. He presented himself as a left-winger who felt that the Left had lost its way. The siren song of French philosophy and literary theory had led the Left, and progressive politics more generally, away from its earlier alliance with science and landed it in a useless and pretentious quagmire.

Many philosophers in the English-speaking world felt vindicated by the Sokal hoax. Although English-speaking philosophy had produced radical ideas about science, for the most part it had not accepted postmodernism and other French-influenced literary-philosophical movements. Jacques Derrida, perhaps the most famous figure in all the humanities during this period, had never been embraced by the philosophical establishment and was regarded by many as practically a charlatan. Some mainstream philosophers of science, who had been made to look dried up and boring by decades of racier work in neighboring fields, were elated. At the 1996 meetings of the Philosophy of Science Association, the presidential address was given by Abner Shimony, a philosopher of physics. Shimony's address was a reassertion of Enlightenment values, the values of science, democracy, rationality, equality, and secularism. Shimony called Sokal "a hero of the enlightenment" for his work in unmasking the foolishness of radical science studies.

Although some philosophers felt vindicated, others felt that damage had been done. After years spent bridging gaps between disciplines and establishing dialogue, Sokal's work was likely to polarize everything again (this argument was made in the discussion after Shimony's talk by Philip Kitcher and Arthur Fine). The fear was that philosophers would cease once again to pay attention to work in neighboring fields. Sociologists, on the other side, would think that the underlying conservatism of philosophy had been revealed again; the smug philosophers had sided with Sokal's cheap shot.

Science studies was not seriously damaged by the Sokal hoax, but there have been some lasting effects. There is less tolerance now for very jargon-laden and obscure writing. That is a good thing, and reason enough to be glad of what Sokal did.

At the start of this chapter I described a shift, during the latter part of the twentieth century, in how people thought about the political role of science. Simplifying quite a lot, science went from being seen as a generally liberating force to being seen as an institution that helps maintain political inequalities of several kinds. Since that time, attitudes have shifted again. In chapter 7 I noted that Bruno Latour, the mischievous sociologist of *Laboratory Life*, had moved to offering explicit defenses of science, motivated by the need to take climate change seriously. For the same reason, a similar renewed friendliness to science is seen in many activists on the "progressive" side of politics. One the other side, some pro-business conservatives, seeing anxiety about climate change as leading to excessive interference in industry, try to emphasize the uncertainties in scientific models of our effects on the climate. This landscape is not static. By the time you read this chapter, it might have shifted again.

8.6 *Values in Science*

In this last section of this chapter I will step back from the details of debates about science and politics, and look at some general themes around the interaction between science and values.

Here is a question that is often asked: should science be value-free? There is also a standard way of describing the debate that ensues. From one side, the answer is *yes*: a crucial part of the intellectual style developed in the Scientific Revolution is the dispassionate and disinterested study of nature, and attempting to be guided by data without prejudices of any kind. The values and preferences of a researcher might sometimes intrude, but this is always something to regret and avoid. When values do have an effect, they make science less objective. The answer from the other side of this standard debate is *no*: it would be impossible for science to be value-free. This is presented as part of the message of Kuhn and the sociology of science. Given this, it's best to accept the situation—accept the "impurity" of scientific choices in this respect—and do a better job with it. It's best to allow an integration of scientific decision-making with broader priorities and values, and do so with our eyes open.

I think this whole discussion is being badly set up, and the options misdescribed. Starting again from the basics: science is a human activity, and all human activities are guided by values of some sort. But this might be a desire to understand, a desire to resolve questions about how the world works. If so, that is not "value-free." A particular goal is being pursued because we value understanding. The ideal of an absence of values in science is really not an issue; it is all a question of what kinds of values, and what kinds of goals, are in play.

A person invested in the more standard science-and-values debate might reply: yes, an absence of a scientific role for values is not an option, but there is still a problem here. The problem has to do with the role of values that might be called practical, moral, or political, as opposed to goals like understanding and truth. We can distinguish between *epistemic* and *nonepistemic* values, and between epistemic and nonepistemic goals (where the word "epistemic" refers to knowledge and evidence). Understanding the world is an epistemic value. Protecting the environment, or advancing the power of your nation over other nations, is a nonepistemic goal. And once the question is asked again—is science affected by nonepistemic values and goals, and should it be?—quite a few of people would say something very similar to what they said in the first round above. It would be impossible for science to be unaffected by the political values of scientists, their funders, and other social groups. Given that fact, it's best to accept the situation—to accept the impurity of scientific choices in this respect—and do a better job with it.

I will argue against that view, too. A good way to approach the position I want to defend is by looking at a very old article, from the 1950s, that posed the questions in an especially clear and sharp way. The article, by Richard Rudner, was called "The Scientist qua Scientist Makes Value Judgments" (1953). The Latin "qua" is used here to mean that the scientist *in his or her role as scientist* makes value judgments. Here is what Rudner says.

Since no scientific hypothesis is ever completely verified, in accepting a hypothesis the scientist must make the decision that the evidence is *sufficiently* strong or that the probability is *sufficiently* high to warrant the acceptance of the hypothesis. Obviously our decision regarding the evidence and respecting how strong is "strong enough," is going to be a function of the *importance*, in the typically ethical sense, of making a mistake in accepting or rejecting the hypothesis. Thus, to take a crude but easily manageable example, if the hypothesis under consideration were to the effect that a toxic ingredient of a drug was not present in lethal quantity, we would require a relatively high degree of confirmation or confidence before accepting the hypothesis—for the consequences of making a mistake here are exceedingly grave by our moral standards. On the other hand, if say, our hypothesis stated that, on the basis of a sample, a certain lot of machine stamped belt buckles was not defective, the degree of confidence we should require would be relatively not so high. *How sure we need to be before we accept a hypothesis will depend on how serious a mistake would be.*

The examples I have chosen are from scientific inferences in industrial quality control. But the point is clearly quite general in application. (1953, 2)

As Rudner notes, this is compatible with the view W. V. Quine was arguing for around the same time, his "web of belief" view, discussed in chapter 2. And it is very common indeed for some hypotheses to be given the "benefit of the doubt" in some way. In statistics, when there are several options on the table, it is common to label one option as the "null hypothesis." This term has several meanings. It can mean the hypothesis that nothing much is going on—"null" as in nothing. But in other cases the null is said to be the hypothesis that we just want to make sure we don't wrongly reject, the hypothesis we want to be sure we won't reject in a situation where it is true. The "precautionary principle," which is often invoked in discussions of environmental and health policy, also fits, in some versions, with what Rudner is saying. That principle is sometimes expressed like this: we do not need to be *certain* that a pesticide (for example) will cause harm before concluding that it is not safe. A lower standard of evidence can be sufficient.

A problem that affects a great deal of discussion in this area is clearly illustrated by the Rudner passage I quoted. Rudner talks about *acceptance* of a hypothesis. This is a decision to treat a hypothesis as true. His picture is that the evidence pushes us a certain distance, but can't take us all the way. One has to go beyond the evidence and decide whether to accept a theory or claim. This is like a (small) leap, or at least an extra move, one that goes beyond merely noting where the evidence pushes.

If we have to make a further decision of this kind in order to "accept," something, how do we do it? As the evidence has already taken us as far as it can, it seems that what is left is the practical side; as Rudner put it, *"How sure we need to be before we accept a hypothesis will depend on how serious a mistake would be."* The appearance of a problem here—a gap that has to be bridged, between evidence and acceptance—is a consequence of thinking about acceptance, and belief, in a particular way. It assumes an "on-off" treatment of acceptance. You either accept a view or you don't.

Suppose we work within a different framework, one that sees belief as a matter of degree. Imagine belief as measured on a scale from 0 to 1. (Or 0 percent to 100 percent.) The number measures confidence— how sure you are of something. You might be fairly sure, very sure, or completely sure.

If we work within a view like this, there is no problem of the kind Rudner wanted to solve (Jeffrey 1956). You look at the evidence, and come to have some degree of confidence in a theory. You might think it is pretty likely to be true, or very likely—whatever. You don't have to then take an extra step and ask whether you "accept it." Degrees of belief can then be used to make practical decisions. You might be fairly confident that it will rain today. Will you take an umbrella? This will depend on how annoying it would be to carry an umbrella around all day if it does not rain, and how annoying it would be not to have an umbrella with you if it does rain. If you are 80 percent confident it will rain, then, given those other factors, you might bring the umbrella, but not if you are only 60 percent confident. I'll go through a model of this sort of reasoning in detail in a later chapter when we look at degrees of belief more closely. For now, I just want to introduce the idea and show that if we use this sort of framework, the kind of decision that Rudner is talking about is just not needed. (I am assuming here that observational evidence can shape your degree of belief in a scientific theory—that will also be discussed later when we return to theories of evidence.)

If we have degrees of belief and can use them to guide action, there is no need to take the extra step that Rudner wants us to. But perhaps we still could? Might we still allow his extra step, and give the practical side of things a further role in some cases? Not only is this unnecessary, it is also a bad idea.

On the view I am sketching, there are two steps in how we handle situations of the sort Rudner is describing. First, we work out what to believe—we work out how confident we are that various hypotheses are true. Then we work out what to do, in light of our beliefs. Our decisions about actions make use of our degrees of belief, but not vice versa. Going back to the umbrella example, we don't need to work out whether to accept that it will rain—in a leap beyond the evidence—to work out whether to carry the umbrella. And it would be a bad idea to do this, because usually there will be a range of different decisions about action that any one belief will be relevant to. If you think it might rain, this can affect decisions about carrying umbrellas, gardening, sporting events, and various other things. Rudner said that how sure we need to be before we accept a hypothesis will depend on how serious a mistake would be, but there are all sorts of different mistakes possible in the different contexts in which we act. Which of these will be allowed to affect whether or not we "accept" a hypothesis? Also, we usually don't know what sorts of actions might be influenced by a particular belief; new choices can arise. The way to handle these situations is to have a degree of belief that is not affected by *any* of the practical decisions on the table, but only by evidence. The degree of belief is updated as new evidence comes in, and whenever we have to act, we make use of the degree of belief we currently have, along with our views about the goodness and badness of the various outcomes that might come from our action.

Again, the idea of degree of belief will be described in more detail later. What I am using it to do in this first discussion is remove the appearance of a gap between evidence and acceptance of the sort that Rudner argues for, and also to argue that it's good to keep a separation between decisions about what to believe and decisions about what to do. Working out whether to ban a pesticide, whether to require that children be vaccinated, whether to stop using coal as an energy source, whether to fund research on human genetic differences, whether to take the time to investigate a possible minor source of error in a piece of work, whether to give a lot of time to unorthodox opinions in scientific debates . . . these are all decisions about what to do. Deciding whether to write a paper is also a decision about what to do, as is working out what to say in a conversation. Many of these decisions are difficult; working out the costs of various kinds of possible error is often not easy. Working out what we think is true is one kind of input to decisions about what to do, along with the risks and benefits that derive from different actions. What I am against, especially in this scientific context, is views that deliberately mix the two together—working out what to believe and working out what to do.

Further Reading and Notes

Keller and Longino's *Feminism and Science* (1996) and Kourany's *The Gender of Science* (2002) are both useful collections. The latter includes the Hrdy paper I discuss in section 8.3. Hrdy's book *The Woman That Never Evolved* (1999) is a more detailed discussion of her ideas. Donna Haraway's *Primate Visions* (1989) is a detailed historical and sociological discussion of primatology from a feminist point of view. For another interesting case study, see Lloyd (2005) on theories of the evolution of female orgasm. Harding's *The Science Question in Feminism* (1986) and Longino's *Science as Social Knowledge* (1990) are two of the most influential books in feminist epistemology as applied to science.

Biagioli's *The Science Studies Reader* (1999) is a good collection that illustrates the diversity of work in that field. For the science wars, see Gross and Levitt's *Higher Superstition* (1994), which includes criticisms of Bloor, Latour, Shapin, Schaffer, Harding, Longino, and various others I have discussed in the last few chapters. See also Koertge's *A House Built on Sand* (1998). The Sokal hoax is the subject of a book of that name edited by the *Lingua Franca* editors (2000). There is also a mass of material about the Sokal hoax on the Internet (starting with website [4]). For modernism and logical positivism, see the Galison paper cited also in chapter 2, "Aufbau/Bauhaus" (1990). Shimony's presidential address praising Sokal was later published (1996).

Section 8.6 criticized views arguing that practical considerations can properly affect our decisions to accept hypotheses, because of the consequences of these choices for our actions. This debate is ongoing. Hempel (1965) introduced the term "inductive risk" for the practical risks that stem from the possibility of error in an inductive argument, and this term is now commonly used. Douglas (2009) is an influential recent discussion. Elliott and Richards (2017) is a collection of papers exploring the problem through case studies.

Another literature looks at whether moral considerations can affect the justification of beliefs in a way that does not depend on actions (Gendler 2011; Basu 2018; Moss 2018). Is making inferences using racial stereotypes, for example, wrong even when practical decisions are not at stake, and even when the stereotype has some statistical value, given prevailing social circumstances? In this literature, "evidentialism" is the view that only considerations that are evidence-related should affect our beliefs (or our degrees of belief). I favor evidentialism, though this literature raises interesting questions. The position I argue for in section 8.6 is evidentialist. Rinard (2019) argues against evidentialism in general. Kincaid, Dupré, and Wylie (2007) is a wide-ranging collection of papers about values in science.

Chapter 9

Naturalistic Philosophy

9.1 *What Is Naturalism?*

What kind of theory should the philosophy of science try to develop? For the logical empiricists, the philosophy of science is concerned above all with the logic of science. By the middle of the 1970s, this view had well and truly broken down. Many wondered whether philosophy had become desiccated and irrelevant. As we saw in some previous chapters, this led to attempts by other fields to annex some of the traditional territory of philosophy of science. If philosophers could not say anything useful about how science works, others would do it instead.

Most philosophers came to agree that philosophy of science had to go beyond logical analysis, and they began, in different ways, to develop alternative views about what sort of activity philosophy should be. In this chapter I will look at one approach that has become influential over recent decades: *naturalism*.

Naturalism is often summarized by saying that "philosophy should be continuous with science." The slogan sounds good, but it is hard to work out what it really means. Naturalists reject the idea that philosophy should be sharply separated from other fields, and think there should be some kind of close connection between scientific theories and philosophical theories. But they do not agree on what this connection should be like. What does a naturalistic outlook on philosophy mean in practice? In this chapter I will first describe naturalism in general and then illustrate the approach with examples. From this point onward, the book also starts to depart from the chronological structure that guided earlier chapters. The remainder of the book is organized more by topic than by chronology.

A moment ago I said that naturalists think that philosophy should be continuous with science. Perhaps a better summary is that philosophy can *use* results from the sciences to help answer philosophical questions, and can do this even in the philosophy of science itself.

From the perspective of many other philosophical positions, to use scientific ideas when theorizing about science involves a vicious circularity. How can we assume, at the outset, the reliability of the scientific ideas that we are trying to investigate and assess? Surely we have to stand outside of science when we are trying to describe its most general features and assess its methods?

The idea that the philosophy of science should work from an external and more secure standpoint is often referred to as *foundationalism*. (This term is sometimes used for other ideas as well.) Foundationalism requires that no assumptions be made about the accuracy of particular scientific ideas when doing philosophy of science. This is because before our philosophical theory is established, the status of scientific work is in doubt. Naturalism is opposed to foundationalism in philosophy.

Naturalists think that the project of trying to give general philosophical foundations for science is doomed to fail. They also think that a philosophical "foundation" is not something science needs, in any case. Instead, we can only hope to develop an adequate description of how knowledge and science work if we draw on scientific ideas as we go. The description of knowledge and science that results will be no more certain or secure than the scientific theories themselves.

Most philosophers who call themselves naturalists would agree with that sketch. From that point on, there is disagreement. "Naturalism" is one of those words that a wide range of people find appealing as a label. Philosophers, like shampoo manufacturers, would always like to call their products "natural." There is a risk that naturalism as a movement will be swamped by the overuse of the term and will dissolve into platitudes. Despite this risk, "naturalism" is the label I use for much of my own work; this is the camp in contemporary philosophy that I generally support, though I will look at reservations and qualifications that I think are reasonable, too.

9.2 *Quine and Others*

The birth of modern naturalism is often said to be the publication of W. V. Quine's paper "Epistemology Naturalized" (1969). But modern naturalism did not come entirely out of Quine. The American philosopher John Dewey is usually thought of as a pragmatist, but during the later part of his career (from roughly 1925 onward) he presented his philosophy as a form of naturalism. In some areas, Dewey's version of naturalism is superior to Quine's. But Dewey's philosophy was neglected during the second half of the twentieth century, and Quine is the figure who had the most influence on the movement.

In "Epistemology Naturalized," Quine made a number of claims. He first attacked the idea that philosophers should give foundations for scientific knowledge. Quine's claims on this point have become widely accepted by naturalists. But Quine also made a more radical claim. He suggested that epistemological questions—philosophical questions about evidence and justification—are so closely tied to questions in scientific psychology that epistemology should not survive as a distinct field at all. Instead, epistemology should be absorbed into psychology. The only questions asked by epistemologists that have real importance, in Quine's view, are questions that can be answered by psychology itself. Psychology will eventually give us a purely scientific description of how beliefs are formed and how they change, and we should ask for no more.

This version of naturalism is one that I, and many others, oppose. Quine seems to be claiming that philosophers interested in questions about belief and knowledge should close up shop and go home. It's not surprising that philosophers, especially those who want to remain employed, would object to Quine on this point, but it's not only a matter of wanting to keep paying the rent; there is a deeper issue here.

In a different version of naturalism, there is such a thing as a *philosophical question*, distinct from the kinds of questions asked by scientists. A naturalist can think that science can contribute to the answers to philosophical questions, without thinking that science should replace philosophical questions with scientific ones. That is the version of naturalism that I defend. This contrasts with the kind of naturalism described by Quine in his 1969 paper; there we think of science as the only proper source of questions as well as the source of answers.

If we think that philosophical questions are important and also tend to differ from those asked by scientists, there is no reason to expect a replacement of epistemology by psychology and other sciences. Science is a resource for philosophy, not a replacement.

What might be examples of these questions that remain relevant in naturalistic philosophy but are not directly addressed within science itself? Many naturalists have argued that *normative* questions are important cases here—questions that involve a value judgment. If epistemology was absorbed by psychology, we might get a good description of how beliefs are actually formed, but apparently we would not be told which belief-forming mechanisms are good and which are bad. We would not be able to address the epistemological questions that have to do with how we *should* handle evidence, and how to tell a good argument from a bad one. Those questions are central to philosophy. For the naturalist, the answers to these questions will often depend, to some extent, on facts about psychological mechanisms and the connections that exist between our minds and the world at large. But the naturalist expects that it will remain the task of philosophy to actually try to answer these questions.

The term "normative naturalism" is sometimes used for naturalistic views that want to retain the normative side of epistemology. I should also note that although Quine's original discussions seemed to leave no place for normative questions in epistemology, toward the end of his career he modified his view, bringing it closer to normative naturalism (Quine 1990). Normative naturalism accepts many (though not all) of the normative questions that have been passed down to us from traditional epistemology. But what is the basis for making these value judgments? What is the basis for a distinction between good and bad policies for forming beliefs? At this point, normative naturalism confronts the old and difficult problem of locating values in the world of facts.

In the face of this problem, some normative naturalists have chosen a simple reply. The value judgments relevant to epistemology are made in an *instrumental* way. In philosophical discussions of decision-making, an action is said to be *instrumentally rational* if it is a good way of achieving the goals that the agent is pursuing, whatever those goals might be. When assessing actions according to their instrumental rationality, we do not worry about where goals come from or whether they are appropriate goals. We just ask whether the action is likely to achieve outcomes that the agent desires. And if some action *A* is being used as a means to *B*, then it is a factual matter whether or not *A* is likely to lead the agent to *B*.

It is uncontroversial that *one* kind of rationality is instrumental rationality. These kinds of norms or values are so closely related to ordinary factual matters that they don't raise much of a problem. It is more controversial to say that this is the *only* kind of rationality. Some naturalists think that instrumental rationality is the only kind of rationality that is relevant to epistemology, while others would like to recognize something further. Can we have meaningful debates about which goals are the right goals, as well as how to achieve a given set of goals? Are our basic goals (in science, or in life) just a matter of personal preference? Are there values in this area that are quite separate from the pursuit of goals? Those are difficult questions, but that is not a reason to simply give up on them. And they are philosophical questions—we can't expect them to be directly answered by science.

Another kind of question asked by philosophers has to do with the relationships between our common-sense or everyday view of the world, on one hand, and the scientific picture of the world, on the other. What kind of match (or mismatch) is there between the two pictures? We find questions of this kind in epistemology: what relationship is there between the common-sense or everyday picture of human knowledge and a scientific description of how it works?

Yet another set of philosophical questions a naturalist can recognize concerns relations between different sciences. Each of the particular sciences gives us fragments of a picture of what the world is like and how it runs. Do the fragments tend to fit together neatly, or are there mismatches and tensions between them? The philosopher patrols the relationships between adjacent sciences, occasionally climbing into a helicopter to get a synoptic view of how all the pieces fit together. This can result in philosophical criticism of particular scientific ideas, but the criticism is made from the point of view of an overall picture that is based in science. The American philosopher Wilfrid Sellars gave a famous one-line definition of philosophy in 1962. He said philosophy is concerned with "how things in the broadest possible sense of the term hang together in the broadest possible sense of the term." Philosophy aims at giving us an overall picture of what the world is like and how we fit into it. This is a good summary of philosophy, and I think it pushes toward a sort of naturalism.

I generally support a naturalistic approach of the kind described in the last few pages, but I don't argue that all work done by philosophers should be naturalistic in this sense. My overall view of philosophy is more open-ended than this. I think philosophy benefits from its unconstrained, unpredictable character. One never knows where the next interesting idea will come from, and when naturalists try to rule out some kinds of work, as they occasionally do, I resist this. Even if we are especially concerned with science, the inherent adventurousness of philosophy plays a valuable role here, too. Philosophy has often served as an "incubator" for novel, speculative ideas, giving them room to develop to a point where they may become scientifically useful. When philosophy is playing this role, it is good not to be too much guided by the science of the day, because the goal is to depart from current pictures and explore new territory.

Michael Friedman has discussed a number of examples of this phenomenon—situations where philosophy of adventurous kinds has fed into science in a way that could not have been anticipated. One example concerns the development of the physics of matter and fields in the nineteenth century (Friedman 2007). Immanuel Kant (someone we encounter a number of times in this book) used his theories about thought and knowledge to offer a view of what matter had to be like. He argued that rather than thinking of matter as simply a *stuff*, we should understand matter in terms of forces of attraction and repulsion ([1786] 2002). These are the forces that, when operating in a certain pattern, give rise to the goings-on that we associate with matter—such as the way one hard object resists being squashed into the same space as another. This is known as Kant's "dynamical" view of matter—a view of matter based on what it does. After Kant, some of the idealist philosophers influenced by him extended this way of thinking in adventurous directions, looking for ways of unifying chemical and physical phenomena, for example. They may have influenced the experimentalist Michael Faraday (via the poet Samuel Taylor Coleridge), and they definitely influenced Hans Christian

Ørsted, who in 1820 was the first person to observe that an electrical current in a wire can give rise to a magnetic field, deflecting a compass needle. The philosophical work had freed up thinking about matter, and suggested the possibility of hidden connections between apparently distinct aspects of nature. This case was surprising to me, as work in that idealist tradition is often seen as wildly speculative and ungrounded, especially when it touched on empirical questions, and very far from responsible science. So it was, but it suggested new ways of thinking that led somewhere important.

If I value this more exploratory kind of philosophy, why did I say earlier that I generally support naturalism? The answer is that I see this incubator role as somewhat secondary in comparison with the role for philosophy expressed by Sellars. Also, work in the exploratory mode will generally not take philosophy back toward the main ideas that naturalism opposes, such as foundationalism. Ruth Millikan (2017) has expressed the combination of stances that I am describing here by saying that the philosopher should embrace a role in which one tries to be partly *behind* science and partly *before* it: behind it when putting pieces of existing science together into an overall picture, and before when pushing ahead with new possibilities.

9.3 *The Role of Observation in Science*

In the rest of this chapter, I will look at two areas that illustrate the naturalistic approach to philosophy of science. One is more individualistic, the other more social. The first is the role of observation in science. In the empiricist tradition, observation (or sensory experience more broadly) is seen as the sole source of knowledge about the world. Observation is the means by which we should arbitrate disagreements, the way we learn which theory to prefer. One of the most important themes of anti-empiricist work as it developed from the 1960s onward was an attack on empiricist views about observation, especially the idea that observation can be an unbiased or neutral source of information when choosing between theories. Instead, it was argued that observations tend to be "contaminated" by theoretical assumptions. This is called the debate about the *theory-ladenness of observation*. Advocates of radical views of science, of the kind discussed in the last few chapters, have often seen the theory-ladenness of observation as a powerful argument against empiricism. The reason the debate is discussed now, in this chapter, is that the debate becomes much easier to settle if it is approached from a naturalistic point of view. The issue gives us a good illustration of how naturalistic philosophy can work in practice.

Arguments for the theory-ladenness of observation were developed by several people around the same time in the mid-twentieth century, especially Kuhn, Feyerabend, and Norwood Hanson. These arguments are a mixture, but their general upshot is supposed to be that observation cannot function as an unbiased way of testing theories (or larger units like paradigms), because observation is affected by the theoretical beliefs of the observer.

I will look at several different ways that theories might affect or contaminate observation. Theories might affect *where you look*; they might affect *how you interpret what you see*; and they might affect *what you see*. These possibilities (all of which might hold) have very different consequences.

First, it is sometimes claimed that observation is guided or affected by theories because theories tell scientists where to look and what to look for. This is true, but no sensible empiricist has ever denied it. This fact does not affect the capacity of observation to act as a test of theory, unless scientists are refusing to look where unfriendly observations might be found. All empiricists would regard that as a breakdown of fundamental scientific procedures.

The second possibility is that theories might affect how you interpret what you see. This, in turn, could work in a number of ways. First, scientists do use theoretical assumptions to decide which observations to take seriously. Some apparent observations involve malfunctions or mistakes of various kinds, and can be disregarded. But it can be hard to tell which ones they might be. Theoretical beliefs can affect this "filtering" of observations, and there is certainly a possibility of bias and other problems there.

This is another manifestation of the holistic nature of testing, which was introduced in chapter 2 and has arisen several times since then. It is not a knockdown argument against empiricist views about observation; the empiricist can reply that the assumptions affecting the relevance of an observation to a piece of theory can themselves be tested separately. We might also venture some recommendations: perhaps in crucial tests, scientists should be more reluctant to discard observations. But in this area it is hard to know which pieces of common sense are helpful, which are trivial, and which are just wrong.

Here is an interesting recent example of what can happen here (for which I am indebted to Adam Collingridge). In human bodies there are two kinds of fat, which are called brown and white adipose tissue. It had long been believed that brown adipose tissue is not found in adults. When new scanning technologies were developed in the first decade of the twenty-first century, scientists began to see things that looked like areas of brown adipose tissues in adults, but initially wrote them off as artifacts of the scanning technology. Eventually, further biochemical tests showed that people had been wrong to blame the machines; the brown tissue was there (Nedergaard, Benson, and Cannon 2007).

Some discussions of this afterward said that people were "not seeing what was in front of their face," and things like that. But that wasn't really the problem. It was a matter of interpretation; they were seeing the marks and interpreting them in a certain way. This shows the separation between some different kinds of influence of theory on observation.

Another argument that involves this second kind of theory-ladenness (the kind that involves interpretation) concerns the role of language. When a scientist has an experience, he or she can only make this experience relevant to science by putting it into words. The vocabulary used, and the meanings of even innocent-looking terms, may be influenced by the scientist's theoretical framework. Given the interconnections between the meanings of words in a language, there is no part of language whose application to phenomena is totally unaffected by theories.

Some versions of this argument are not of enduring importance because they cause trouble only for the early logical positivist ideal of a pure observational language. Sometimes critics of empiricism write as if once it has been shown that the language of observation is in some sense "theoretical," that is the end of the argument and empiricism is dead. This is a mistake. In working out the relevance of this issue to more modern forms of empiricism, everything depends on which *kinds* of theories affect the language of observation and on the *nature* of this effect. For example, maybe observational reports make use of theories that are so low-level that the testing of real scientific theories will never be affected. We can think of the assumption that objects generally retain their shape when we are not looking at them as "theoretical," in a sense, but the effect of this assumption on observation reports does not usually matter to testing theories in science.

Suppose it can be shown that observation reports use language that is closely tied to the kinds of theories that are themselves being tested. For example, Feyerabend tried to show that innocent-looking descriptions of motion in the seventeenth century were affected by theoretical assumptions in this way (chapter 6). This looks like trouble. But even this kind of effect may or may not be a problem. Not every result described in terms of the *concepts* preferred by theory *T* will be an observation report that is *favorable* to theory *T*. Back in my discussion of Popper, I mentioned that an observation of rabbit fossils in Precambrian rocks would be a massive shock to evolutionary theory. Suppose we regard "I saw rabbit fossils in Precambrian rocks in Russia" as an observation report that is very much "laden" with biological and geological theory. Some might want to say it is so laden with theory that it is not an observation report at all. But regardless, the report would still be a massive shock to evolutionary theory. The fact that observation reports are expressed using concepts derived from a theory need not affect the capacity of nature to say *no* to a theory.

The third and final aspect of the theory-ladenness arguments that I will consider is the most important. Kuhn and others have argued that even the experiences themselves that people have are influenced by their beliefs, including their theories. There is *no stage* in the processes of observation in science where theories do not play a role.

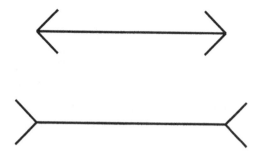

Figure 9.1. The Müller-Lyer illusion

When giving these arguments, Kuhn and others liked to use the results of psychological research in the middle of the twentieth century. This research was taken to refute a "passive" view of perception and replace it with a view holding that perception is active and intelligent. Psychologists emphasized the multiple ways in which a pattern of stimulation on the retina could be caused by objects in the world. If there are multiple possibilities, then theoretical assumptions must be used by the visual system to make a choice (Gregory 1970).

This kind of theory-ladenness argument was attacked by Jerry Fodor in an 1984 paper called "Observation Reconsidered." Fodor turned the tables on some theory-ladenness arguments with a discussion of perceptual illusions of the kind often discussed in psychology textbooks. Consider the Müller-Lyer illusion, represented in figure 9.1. The two lines are the same length, although we tend to see the lower one as longer than the upper one. Roughly speaking, the illusion is created by the unconscious use of background assumptions in the processing of visual inputs. People have taken this result to show a kind of theory-ladenness in perception; though we do not realize it, our beliefs about the world are affecting what we see.

But Fodor then notes that although it is true that the illusions seem to be produced by the effects of unconscious theory, *some* pieces of theory or background knowledge seem to have no effect on perception. Most notably, the illusion is not affected by the knowledge that it is an illusion, or by knowledge of the theory of illusions. Those pieces of background knowledge do not make the illusion go away. The mechanisms of perception seem to be influenced by some theories and not others. And the theories that do have an effect are not sophisticated scientific theories. They are low-level sets of assumptions about the physical layout of the world—the three-dimensional nature of space, the effect of distance on apparent size, and so on. The mechanisms of perception then send information to more "central" mechanisms in our brains, the ones with which we think and make deliberate decisions. When these central mechanisms are at work, then all sorts of other beliefs and assumptions play a role. Here the holistic nature of testing becomes relevant—we might come up with all sorts of explanations for what we seem to be seeing. That is still a problem, but it is not a problem about observation itself.

Note also that our perceptual mechanisms might use low-level theories in a way that makes perception quite reliable, without these theories themselves being true. Our brains might have something like Euclidean geometry built into them, for example. This geometry is not strictly true of our universe, but it might be used by us in such a way that we usually end up with true observational judgments (about what is in front of what, for example). And the Euclidean habits we use for interpreting space in everyday experience did not prevent us from revising the scientific belief that space strictly fits Euclidean geometry, as Einstein did in his theory of general relativity.

It should be clear how this discussion provides support for a naturalistic approach to philosophy. Observation is a natural phenomenon, studied by fields such as psychology and psychophysics. Those disciplines tell us what perceptual mechanisms are like and what kind of connection we have to various parts of the world by means of these mechanisms. They tell us when ordinary human perception is reliable and when it is not. The initial impression we all get in the Müller-Lyre case is, of course, mistaken; it takes some effort to work out what is going on. Naturalistic philosophers can put all these results to use in working out how observation actually operates, and how it could operate, in science. The theory-ladenness of observation is an important philosophical problem—or, more accurately, collection of problems. It affects debates about empiricism. But the way to assess claims about theory-ladenness is to look at various bits of science itself.

A related question about observation was raised in chapter 2: is there much of a difference between observations that come from experiments and those that do not? I'll come back to this question a couple more chapters down the track.

9.4 *Science as a Process*

Traditional empiricism neglected the social structure of science. Kuhn showed how much is lost if you ignore the social side. Naturalistic philosophy has tried to avoid this mistake.

I will discuss a couple of examples, starting with David Hull's work. Hull spent many years observing and interacting with biologists, especially the community of biologists who study *systematics*, the classification of organisms. This led him to some general ideas about scientific communities and how they work, summarized in a sprawling book called *Science as a Process* (1988).

Back in the first chapter, I said that philosophy has a pair of goals in this general area: understanding knowledge in general, and understanding the distinctive features of work that came out of the Scientific Revolution. Hull is looking at the second of these. He takes for granted that humans can perceive and know about the world, not just in scientific contexts but many others. He wants to understand what makes science different.

The story he gives is one based largely on the social organization of incentives and rewards, especially a combination of *cooperation* and *competition*. People sometimes ask whether science is a primarily cooperative enterprise, or a competitive one in which scientists are out for personal advancement. Hull thinks the answer is *both*, and the special features of science result from an interaction between the two.

Hull starts from a picture in which scientists are curious about the world, and he assumes that curiosity is a common human trait. What makes science distinctive is how the energies of different people are organized, especially the relationship between the goals of the individual scientist and the goals of science as a whole. He argues that the main professional motivation for individual scientists is the desire for recognition. And one kind of recognition is most relevant: *use*. Scientists want other scientists to use their work, giving credit when they do so. Hull does not say this, but it seems me that his story is one in which the desire for credit often tends to overshadow the curiosity that gets things rolling.

In his emphasis on credit, Hull is fairly close to the ideas of Robert Merton, as he acknowledges. Merton (see chapter 7 above) argued that recognition is the basic reward in science. There are some differences, though. Hull stresses the desire to have one's ideas used; Merton stresses being recognized as the first to come up with an idea. Often these will go together, but not always. In Hull's view, but not Merton's, if a scientist's version of an idea is used because it is especially convenient or clear, even though it was not the first, that counts as a good outcome for the scientist.

When the two views are compared, it seems to me that Hull's is more accurate on this point. Merton's picture is most applicable to high-profile scientists, and to historical periods when there is lot to do and not too many people doing it. Once a field is mature and has a large community of scientists working on details, the Merton-type rewards become very scarce, and something else must be driving most of the people in the field. That driver might in some cases be simple curiosity, which is probably an underestimated motivation in this context. Many scientists are deeply curious about the systems they study; they just want to get to the bottom of what's going on. But another, which can work alongside curiosity, is something like Hull's desire to have one's work used and cited.

The importance of credit in Hull's sense has become steadily clearer in the years since *Science as a Process*, with the rise of citation analysis and the influence of measures like the "h index." (A researcher has an h index of 12, for example, if he or she has written 12 papers that have each been cited 12 times, and 12 is the highest number for which this is true). The h index was introduced by Jorge Hirsch, a physicist, in 2005, and its use in ranking and assessing researchers has now become routine.

(I suspect his *h* was intended to stand for Hirsch, in a Mertonian manner. It could also stand for Hull.)

The distinctive features of science, as Hull sees them, come from this system of incentives as it operates in a special social context. Each scientist inherits the ideas and methods of his or her field from earlier workers. Individual scientists cannot do anything significant without entering a system of cooperation and trust, without using the work of others. And in order to use the work of others in ways that provide support for what you are doing, you need to give citations. So the desire to do work that is used leads to using and citing the work of others. Scientists trade credit for support, and they do so in the hope that others will do the same for them. This reciprocation is not primarily a matter of goodwill (although that can be a factor); instead, it results from a special kind of self-interest.

Hull argues that the reason fraud in science is treated as so much more serious a crime than theft, even in cases where public well-being is not affected, has to do with these sorts of factors. In a case of theft or plagiarism, the only person harmed is the one stolen from. But when a case of fraud is discovered, all the scientists who used the fraudulent work will find their work on that topic deemed unreliable, and their work will not be used.

A number of philosophers and scientists have been attracted to a picture of science as a dialogue between an imaginative voice and a critical voice, between the speculative and the hardheaded. Popper is an example. The picture is an appealing one, but why should this dialogue actually occur? Hull aims to give us a mechanism. Part of the mechanism is the distribution of roles across different people. In contrast to Popper, Hull argues that there is no need for individual scientists to take a cautious and skeptical attitude toward their *own* work; others will do this for them.

So far we have looked at the consequences of the structure of reward and motivation in science. But *why* do scientists want to have their work used in the way Hull describes? This question is related to another: why has the social structure that Hull describes arisen rarely, across all the different communities that have wanted to understand the world? For

Hull himself, this is somewhat puzzling. He says he assumes that both curiosity and a desire for recognition are fairly basic human motivations. That is true, but curiosity is not encouraged in all cultures. In some religious contexts, curiosity is seen as dangerous or unwelcome. The 1611 King James Bible warned, "Be not curious in unnecessary matters: for more things are shewed unto thee than men understand" (Ecclesiasticus 3:23). The Scientific Revolution made curiosity respectable, and this was a cultural shift (Daston 1995; Ball 2013).

Hull's mechanism is also based on a particular kind of credit, one that involves the ongoing *use* of one's work. We seem to have a situation where a basic human desire for recognition and respect is shaped by the internal culture of science into a desire for something more specific. I said "shaped," but we should expect both some shaping and some selection here; individuals who do not find the scientific reward system satisfying might never finish graduate school.

The reward system found in science was a fairly early invention, but it did have to be devised and worked on. When the French Academy was founded in the seventeenth century, its members initially tried to handle credit communally. This, they found, did not work, and they switched to a more individualistic approach. The Royal Society of London, under its skillful first secretary, Henry Oldenburg, used rapid publication in the *Proceedings* both to allocate individual credit and to encourage people to share their ideas. Oldenburg's system, which also included anonymous refereeing of papers, is basically what has come down to us today.

Before leaving Hull, I will make a quick comparison with Latour, another person who wrote a lot about the role of credit and competition in scientific work. Hull's social account of science is set up with a kind of basic empiricism operating in the background. If your scientific ideas keep running into trouble with observational results, people will tend to recognize that and not use those ideas. This won't always happen, but often it will. For Hull, there is much underlying constraint coming from experiments, predictions, and the like. In Latour's story, that is not how things work. Speaking roughly, we might say that for Latour, the empirical side of things is a resource for scientists, something they can choose to make use of in their debates; for Hull, it is a constraint.

9.5 *The Division of Scientific Labor*

In chapter 6 I discussed Lakatos's and Laudan's views about competition between research programs. Both presented a picture, which seems to cover some fields in science fairly well, based on competition between teams of workers developing rival theories and perhaps defending rival methods. This kind of organization also raises questions about the relations between the goals of individual scientists and the goals of the whole field, questions that Lakatos and Laudan did not address much at all. Some naturalistic philosophy has looked closely at these relationships, using the tools of mathematical modeling.

The first to do this was Philip Kitcher (1990; 1993). He starts by asking this question: suppose you ruled science from on high, and had to allocate resources to rival research programs. In a particular scientific field, you find two different approaches being taken to the same problem. Research program 1 looks more promising than research program 2. No one knows which approach will ultimately work, but it is clear that one will succeed while the other fails, or both will fail. How should you allocate resources if you want to maximize the chance that the scientific problem will be solved?

The answer will depend on the details of the case, but it seems that in a wide variety of situations the best approach will not be to allocate all resources to one option and none to the other. Some degree of "bet-hedging" will often be advisable, even when one program is more promising than the other. A wise "ruler of science" would often allocate most resources to the better option but some to the alternative.

Here is a simple case. Suppose that research programs always become more likely to succeed as more workers are added to them, but there is also a "diminishing marginal return." As more workers are added to a program, each additional worker makes less and less difference to the chances of success. We can then see why an optimal allocation of resources will often not put every worker on one program. After a certain point, adding more workers to a program has almost no effect, and these people should instead be put to work on the alternative. Unless the total pool of workers is small and the overall difference in promise is big, the best distribution of workers will allocate some to one program and some to the other.

Now suppose there is no "wise ruler of science" who can tell people what to do, and individuals have to make their own choices about which program to work on. Then we can ask: what kind of individual reward system in science will tend to produce distributions of workers that benefit science as a whole? What reward system will tend to produce the *same* distribution of workers that a ruler from above, who only cares about the solving of the problem, would want?

One option that would *not* work well would be to give a fixed reward to everyone who works on the program that eventually succeeds, regardless of how many people were working on it. That would induce everyone to choose the more promising program, and the community would have all its eggs in one basket. Another approach would be to reward individuals for making choices that produce the best effect on the chances that the community will solve the problem. This could work in principle, but it does not seem a realistic reward system for actual people. So here is a third option: we reward only the individuals who work on the research program that succeeds, but we divide the pie equally between all the workers who chose that program. The reward that an individual gets will depend not just on their own choice but on how many other individuals chose the same program.

This system will produce a pretty good distribution of workers across the two options. Once one research program becomes crowded, an individual has little incentive to join that program because, if it does pay off, the pie is being divided among too many people. Although the other program is less likely to succeed, if it does, there will be fewer people sharing the reward. An individual who wants to maximize his or her "expected payoff" will often have reason to choose the less promising program. In this way, selfish individual choices will produce a good outcome for the whole community. Kitcher suggests that this reward system is fairly close (with simplifications) to what we actually find in science. The pie is not cash but recognition and prestige.

Kitcher's story has an "invisible hand" structure of the kind I discussed in chapter 5 when we looked at Kuhn. We have selfish individual behaviors combining to produce a good outcome for the community. This outcome might be one that the individuals are uninterested in or even unaware of.

Kitcher's work has been followed up by Michael Strevens, Kevin Zollman, and others. Strevens (2003) argues that Kitcher was too optimistic about the reward system in which a fixed pie is shared equally by all workers on a successful program. Suppose you are making a choice of which program to join. There are cases where you will do best to join the more promising program even though your joining makes little or no difference to its chance of success. Others have given the program a good chance of success, and your joining gives you a good chance of an equal share of the pie, though your efforts would have been more productive if you had joined the alternative program. A kind of "free riding" might be encouraged in Kitcher's scheme.

Strevens argues that another reward scheme is both better for the community *and* closer to the actual situation in science. This scheme allocates rewards to an individual that are proportional to the contribution they make to the research program they join. The payoff is given only if a research program solves the scientific problem, and the pie is shared unequally among those working on the successful program. Workers who joined early and/or made a big difference to the program's chance of success get more than workers who joined late and made little difference. Zollman (2010) adds another dimension. Above we imagined scientists choosing a research program and sticking with it, but scientists can often switch if they get information suggesting that the other side is doing better. (A switch might be expensive in terms of time, and socially difficult, but often it will be possible.) A good community will be one where the appearance of early success in one program does not induce too big a rush of people to join it. That appearance of early success might be misleading. Zollman shows that in some circumstances, a community can do better if information does not move around too readily—if not everyone learns too quickly about one program's appearance of early success. That can be achieved if the social network linking the workers and research programs is not too densely interconnected—if it takes a while for information to flow across the whole network. There is a delicate balance to strike here; there is no point in having people sticking with a research program indefinitely once another program is clearly solving the problem. Information has to get around eventually. A good community will preserve a diversity of approaches for a while, but not forever.

9.6 *More on Competition and the Goals of Science*

This last section is not so much about naturalism, but continues with some of those themes about individual success and the goals of science as a whole.

A lot of work on this topic in philosophy has tended toward a rather positive view of the relationship, seeing an alignment between individual and group interests. Hull's version of this view was especially optimistic. If you seek to have your work used with credit, in a context where this requires using the work of others and giving them credit, you will learn from those around and before you, and try to improve on their work. People will have reasons to check the work of others that they intend to use (and also the work of people whose work clashes with theirs). If you are careless or dishonest, people will find this out, and your work will *not* be used. And so on. Hull acknowledged that the system malfunctions on occasion, but the general situation is a harmonious relationship between the behavior of individual workers and the goals of science as a whole. We might then expect a good deal of progress and efficiency.

Looking at the history of science as a whole since the Scientific Revolution, I think it is reasonable to say that things have gone fairly well. Individuals have been motivated by the desire for credit to do remarkable work, and to work extremely hard, in ways that have led to a great deal of benefit for large numbers of people. But you can say that without saying that everything is as good as it could be. Might science work just as well, or better, with a somewhat different reward system from what we find today?

For example, do we really need the intense and egotistic competition? This is a hard question to answer. I said in section 9.4 that in the seventeenth century there was a brief period that saw different approaches to allocating credit, and the more individualistic English system prevailed. That was one experiment, and a long time ago. Some might argue on more general grounds that nothing other than a very competitive system is likely generate the intense levels of commitment and dedication seen in modern science, but even if this is true, there may be downsides as well.

One of Hull's case studies is interesting in this context. Hull discusses, and extends, work by sociologists on the temperament and leadership styles of successful scientists. The data suggest that an aggressive faith in one's own ideas, the pushiness of a "true believer," can be useful in at least some fields. Hull refers to a sociological study in which a detailed survey of students and colleagues was used to investigate the temperaments of some famous twentieth-century psychologists in the United States. One contrast, between B. F. Skinner and E. C. Tolman, is especially interesting. Both Skinner and Tolman were in the "behaviorist" tradition in psychology; they wanted psychology to be experimental, quantitative, and closely focused on behavior. But Skinner's version of this approach was almost absurdly strict, while Tolman's was more flexible. Tolman was also a modest, open-minded, considerate sort of person; Skinner was dogmatic and pushy. Skinner had much more influence than Tolman. We cannot know for sure what role temperament had in explaining this difference in success, of course, but the data are suggestive (and Hull found similar results in a smaller study of his own).

Let's assume that pushiness and zeal work well for individuals. Does this tend to result in good outcomes for science? In this case, an argument could be made for the opposite view. I think that if Tolman had dominated mid-twentieth-century psychology rather than Skinner, it probably would have been better for the field.

Back near the beginning of chapter 5, I used Skinner's work as an example of a paradigm in Kuhn's sense. His framework dominated academic psychology (without dominating clinical psychology) in the mid-twentieth century, especially in the United States. Later in that chapter I also talked about its overthrow, in an unusual revolution partly prompted by Noam Chomsky's work in the neighboring field of linguistics. Tolman himself had done some work that put pressure on Skinner's view. Tolman believed that animals build "maps" of their environment in their brains (1948), a concept that would not be allowed in Skinner's psychology. But Tolman did not produce a revolution, while Chomsky did. Since then, Tolman's ideas have had a revival, including the "inner map" idea (O'Keefe and Nadel 1978). Chomsky, a central figure in the revolution that deposed Skinner and behaviorism, was (and is) another ultra-confident and tenacious person, who is sure he is right (despite significant changes of mind) and vigorously attacks those who disagree.

It seems to me that the individual benefits of scientific overconfidence and pushiness probably do not match up with benefits to science as a whole. But impressions don't count for much, and it would be good to get more data on this point, difficult though it is.

We should also bear in mind that the culture of science is not fixed and unchangeable. Scientists have usually not hoped to become rich through their work; recognition has been an alternative form of reward. But financial rewards have started to become a far more visible feature of the life of the scientist, especially in areas such as molecular biology and biotechnology. Kuhn thought that the insulation of science from pushes and pulls deriving from external political and economic life was one source of science's strength. Some of the features Hull approved of and emphasized are also now being put under pressure as the sheer pace of science increases. The race to get grants and to get papers published in high-impact journals may now lead some people to cut more corners than they would under a different regime, and perhaps more than they did in the circumstances in which Hull was writing, in the latter part of the last century. I'll come back to this question in the last chapter, when I look at current trends and where we might be heading.

Further Reading and Notes

The history of naturalistic philosophy is discussed in Kitcher (1992). Kornblith's *Naturalizing Epistemology* (1994) is a collection of papers on the topic, including Quine's classics. Dewey's most important naturalistic work is his *Experience and Nature* (1929a). For "normative naturalism," see Laudan (1987). I discuss the relations between Quine and Dewey (and between naturalism and pragmatism) in Godfrey-Smith (2014). Friedman discusses other examples of cases where philosophical ideas incubated scientific developments in his book *Dynamics of Reason* (2001).

Naturalism is sometimes associated not with a view about how to approach philosophy, as I have it here, but with a view about what the world contains. Non-natural things, in some sense, are ruled out. Much of the time, this comes down to a kind of materialism—a denial of souls, gods, and the like. This is a different position from the one outlined in this chapter. Sometimes it is called "ontological" as opposed to "methodological" naturalism.

The theory-ladenness of observation is discussed in the Kuhn and Feyerabend works discussed earlier in this book, and also in Hanson (1958). Fodor (1984) was the subject of a response by Paul Churchland in the journal *Philosophy of Science* (1988). Fodor replied in the same issue.

Kitcher's main work here is *The Advancement of Science* (1993). The model discussed in this chapter is presented in a simpler form in Kitcher (1990). Solomon (2001) defends "social empiricism" in detail, with examples from the history of science. For a general discussion of social structure and epistemology, see Goldman (1999). Whitcomb and Goldman (2011) is a collection of papers about social epistemology, though mostly not about science. Boyer-Kassem, Mayo-Wilson, and Weisberg (2017) is a collection of papers specifically about social structure and science. When David Hull died in 2010, I wrote an obituary (Godfrey-Smith 2010).

Hirsch (2005) began his paper: "For the few scientists who earn a Nobel prize, the impact and relevance of their research is unquestionable. Among the rest of us, how does one quantify the cumulative impact and relevance of an individual's scientific research output?" This is similar to the point I made about the limitations of Merton-style credit, which can only be reasonably sought by early people in a field or by very high fliers.

In my outline of Hull's ideas I did not mention one of his main arguments. Hull tries to describe scientific change as an evolutionary process, in a way derived from biology. Science changes via processes of variation and selection, just as biological populations do. Individual ideas are replicated in something like the way that genes are, and scientific change is a process in which some ideas outcompete others in a struggle for replication. The idea of understanding scientific change by means of an analogy with biological processes of natural selection has been tried out by a number of writers (Toulmin 1972; Campbell 1974; Dennett 1995). As we saw in chapter 4, Popper's view of science also has an analogy with Darwinian evolution, though Popper did not start out with this analogy in mind. The analogy between science and Darwinian evolution is something that people keep coming back to, and it is certainly intriguing, but I don't think it has yielded a lot of new insights so far. There are other parts of culture where the analogy may gain more traction (Godfrey-Smith 2012).

Chapter 10

Scientific
Realism

10.1 *Strange Debates*

What does science try to describe? The world, of course. Which world is that? *Our* world, the world we all live in and interact with. Unless science has made some very surprising mistakes, the world we now live in is a world of electrons, chemical elements, and RNA molecules, among other things. Was the world of a thousand years ago a world of electrons and RNA? Yes, although nobody knew it back then.

But the concept of an electron is the *product* of debates and experiments that took place in a specific historical context. If someone said the word "electron" in 1000 CE, it would have meant nothing—or, at the very least, not what it means now. So how can we say that the world of 1000 CE was a world of electrons and RNA? We cannot; we must instead regard the existence of these things as dependent on our concepts, debates, and negotiations.

For some people, the claims made in the first paragraph above are so obvious that only a tremendously confused person could deny them. The world is one thing, and our ideas about it are another. For some others, the arguments in the second paragraph show that there is something badly wrong with the simple-looking claims in the first paragraph. The idea that our theories describe a real world that exists wholly independently of thought and perception is a mistake, and one linked to other mistakes—about the history of science, progress, and the trust and authority we should accord science today.

These problems have arisen several times in this book—in chapter 5 we looked at Kuhn's claim that when paradigms change, the world changes too, and in chapter 7 when we found Latour suggesting that nature is the "product" of the settlement of scientific controversy. I criticized those claims, but now it is time to have a close look at these problems.

There is a standard way of setting up the main debate in this area, which runs like this. We ask: does science tell us (at least sometimes, when things go well) about the hidden nature of a world that exists in a mind-independent way? *Scientific realists* say yes. Various opponents of scientific realism say no.

A "no" answer can be reached in a number of ways. Perhaps theories can never "reach beyond" experience in what they say. Perhaps theories can make claims that reach further, but we can't ever expect to get claims of that kind right. Or perhaps the world itself does not exist mind-independently. Views that oppose the "realist" picture are often quite different from one another. I am basically on the realist side in these debates. Certainly I oppose the opposition—all the views that are seen as the main rivals of scientific realism, I oppose. But I've come to think that many parts of the discussion are unsatisfactory, including a lot of what people say on the realist side. There is a collection of problems here, and it is best, I think, to address them one by one, and not to worry too much about whether the view you end up with is a version of "realism."

10.2 *Realism*

The term "realism" has no single meaning in philosophy. Sometimes realism about something—moral values, for example, or God—just means commitment to the existence of that controversial thing. Then the realism debate is a debate about God, or moral values, or whatever the particular topic might be.

There is also a tradition of debate about what our basic attitude should be toward the world that we seem to inhabit—the ordinary world of furniture and trees and so on. The familiar, common-sense view is that this world is out there around us, existing regardless of what we think about it. We are all parts of a single world that exists in space and time, learning about this world when things go well, and acting on it in various ways. That view is sometimes called "common-sense realism."

Despite perhaps being part of common sense, this view has been challenged repeatedly. One kind of challenge holds that we could never *know* anything about a world of that kind, so it makes no sense to think of the world of our everyday dealings as a world that exists mind-independently. We have to somehow bring the world of everyday life "closer in" to the mind. In reply, you might try to show that we can learn about a mind-independent world, or argue that realism (in this sense) can be true even if knowledge of the world is limited or impossible; you might try to be a "skeptical realist."

The usual way to express common-sense realism is by saying that reality exists "independently" of thought and language, independently of what we think and say about it (Devitt 1997). I used that language above. But "independent" is too broad a term. People's thoughts and words are parts of the world, not extra things that somehow are outside or above it. When thoughts change, the world changes, because thoughts are part of the world. And thought and language have an important causal role in the world, by means of action. One of the main reasons for thinking, talking, and theorizing is to work out what to do, especially how to modify and transform things around us. Every bridge or light bulb is an example of this phenomenon. So it would not make sense to just say that realism asserts that "the world" exists "independently" of thought. But still, there is something worth saying here once we have made some qualifications. We could say it like this.

Common-sense Realism: We all inhabit a common reality, which has a structure that exists independently of what people think and say about it, except insofar as reality is comprised of, or is causally affected by, thoughts, theories, and other symbols.

A realist, in this sense, accepts that we may all have different views about the world and different perspectives on it. Despite that, we are all here living in and interacting with the same world. The world I interact with is the world you interact with. We are both parts of the same world, and we can both affect it. This is a view in which the idea that a person's own perspective determines their own world is rejected. It's not true that "we each have our own reality," even though we might each have our own views about what is real.

I think that realism in this sense is right; the arguments against it are unconvincing. Let's now look at how these sorts of questions relate to the philosophy of science.

10.3 *Approaching Scientific Realism*

Suppose someone accepts a realist view in the sense described above. What role does science have? It would be easy and uncontroversial to add that science provides new ways for us to interact with various parts of the world, and it creates new tools and technologies. A realist will see the activity of science as part of the overall picture. The view called "scientific realism" is usually seen as saying more than this, though. One option is to see the scientific realist as asserting that the world *really is* the way it is described by our current and best-established scientific theories. We might say: there really are electrons, chemical elements, genes, and so on. The world as described by science is the real world—for the most part, anyway.

Others, such as Bas van Fraassen, argue that it is a mistake to express the scientific realist position in a way that depends on the accuracy of our current scientific theories. Scientific realism is a philosophical view about science as a whole that does not depend on what our current theories happen to say. If our current theories turn out to be false, we have made some scientific errors, but we might not have made any philosophical mistakes about how science works and what it means when things go well.

In response to this, the instinct of quite a lot of philosophers has been to qualify the claim made by a scientific realist—make it more cautious—but hang onto the basic idea above (Devitt 1997). We might say that scientific realism holds that *most of our mature scientific theories are at least approximately true.* "Mature" theories means those that have been carefully studied and survived a lot of testing. A mature theory in this sense might still be wrong, but then, realists say, its success would be a "coincidence" or a "miracle" (Smart 1963; Putnam 1975a). This is now called the "no miracles" argument—scientific realism is said to be the only view that does not make the success of science in prediction into a miracle. The kind of realism that the no-miracles argument is supposed to support is something like the claim that most of our mature scientific theories are at least approximately true.

Another option is to allow that a scientific realist can be much more pessimistic about whether we are getting things right, and perhaps whether we are likely to ever get things right. We might see scientific realism as a claim about the *aims* of science. There is a world we live in (as in common-sense realism), and science aims to describe it. That is basically how van Fraassen, who I mentioned above and will also return to later, thinks of the scientific realist position. Science might fail, but there is a real world whose structure we are trying to describe with science.

This is a "safer" kind of scientific realism, but it doesn't seem to be saying very much. People can have whatever aims they want. Is there such a thing as *the* aim of science, or even a list of aims? It is probably true that most scientists see themselves as trying to describe how the world really works, but that is just a matter of their attitudes.

We might then say that for scientific realism it's a *reasonable* aim of science to describe or represent the world as it really is. (That is how I did it in the first edition of this book.) The realist says there is a world our work is aimed at, and the tool kit with which we approach it— theories, mathematical models, observational testing, and so on—is one that could work if things go well. It could enable us to describe what is really going on.

That might sound trivial, but plenty of people have argued that there is a big obstacle, of some sort, to our describing how the world works. These people think the scientific tool kit is good enough for some purposes, but not for that one.

I will come back to those arguments in a moment, but first I want to make a broader comment about the relationships between a "realist" view of the world, on one side, and these questions about what attitude we should have toward theories that have been well tested, on the other.

One view that usually gets called a "realist" option is the idea that when a theory has been tested extensively, we have reason to believe what it says (most mature theories are at least approximately true). Something also usually seen as a realist position is the idea that we inhabit a world that exists mind-independently (in the qualified way spelled out in the previous section). However, there is a problem here, at least in principle. One of the things that science might tell us about is whether we live in a world of that kind—a world that exists mind-independently—or not. In some interpretations, quantum mechanics shows us that we do not. According to some versions of this theory, the state of a physical system is partially determined by the act of measurement, in a still-unresolved sense of the term "measurement." A number of physicists have seen this as causing problems for common-sense realist ideas about the relation between human thought and physical reality.

Some people think this is just impossible; it can't be a scientific fact that there are no facts. This is one of the situations where the word "fact" causes problems, though (as it did in chapter 7—we will see a few of these). It can't be a true scientific sentence that there are no true sentences. That would make no sense. But it might be a true sentence, established by science, that the view of the world as containing mind-independent objects is wrong. That is, in principle, possible.

These interpretations of quantum mechanics are controversial, and my understanding is that they are fading in influence. Philosophers of physics tend to be more critical of these views than physicists themselves (e.g., Maudlin 2019). Even if those views are completely mistaken, that is not the point here. The point is that they show there is a possibility that science could conflict with common-sense realism, at least in some versions. So if we were to believe a theory of this kind—if we took a "realist" attitude to the theory—then we would have to drop "realism" in the other sense, the sense that involves mind independence. This is an illustration of problems with the whole idea of scientific realism as it is often understood. If realism combines a claim about the nature of the world we live in with a recommendation that we believe what science says, then there is the possibility of a clash between those two commitments.

I'll introduce another problem, too, this one about the idea that scientific realism involves a sort of overall confidence in the messages of science, such as the view that it would be a miracle if most of our well-tested theories were false. The problem is that different scientific fields might require very different treatments in this area. In a lot of discussions of scientific realism, people look for a general summary of how confident we should be in "science," or "current science," or "mature theories"—where this applies across all the different fields, from physics through chemistry to biology and psychology and the social sciences, and also paleontology and meteorology and other interesting special cases. But we might have reason to have different levels of confidence, and also different *kinds* of confidence, in different parts of science. Ernan McMullin (1984) argued that we should not think of the parts of physics that deal with the ultimate structure of reality as a model for all of science. Physics is where we deal with the most inaccessible entities, those furthest from the domain our minds are adapted to dealing with. In physics we often find ourselves with powerful mathematical formalisms that are hard to interpret. These facts give us grounds for caution even when theories are empirically successful. Those special features of physics do not apply at all in the case of molecular biology. There we deal with entities that are far from the lowest levels, entities that we have a variety of kinds of access to. We do not find ourselves with powerful mathematical formalisms that are hard to interpret. Trying to work out the right attitude to have toward biology is not the same as trying to work out the right attitude toward theoretical physics.

In cases where theories are doing well, we might also have grounds for optimism about some features of a theory and not others. McMullin, John Worrall (1989), and others have developed versions of the idea that the confidence we should have about basic physics is confidence that some *structural* features of the world have been captured reliably by our models and equations. That is a special kind of confidence. There have been interesting cases in the past where a theory had the right structure even though it was, in many ways, quite wrong about the kinds of things that exist. Here is an example used by Laudan: Sadi Carnot thought that heat was a fluid, but he worked out some of the basic ideas of thermodynamics accurately despite this. The flow of a fluid was similar enough to patterns in the transfer of kinetic energy between molecules for his mistake not to matter too much. It's possible to believe that some parts of science are very good at describing structure—relationships between things—without being so good at describing the objects that stand in these relationships. This view, again, probably fits better with physics than other parts of science.

The issues in the last couple of pages should give some indication of why I think that that standard ways of discussing scientific realism have problems. However, you might reply that the issues I talked about just above are a mostly a matter of fine-tuning a basically realist outlook. Perhaps so, and given this, it makes sense next to look at views that depart more dramatically from the picture I've been working with here. Indeed, many of the philosophers who have figured in earlier chapters of this book have seen part of their project as arguing for alternatives to scientific realism. As well as Kuhn and Latour, who I've mentioned already in this chapter, this can be said, in different ways, of Schlick, Goodman, Laudan, and others. So let's look at some rival views.

10.4 *Challenges from Empiricism*

Scientific realism has often been challenged by some forms of empiricism. One part of the debate about realism is often referred to as a debate between realism and empiricism, though I think this is not an inevitable opposition.

Traditional empiricists tend to worry about both common-sense and scientific realism, often for reasons having to do with knowledge. If there was a real world existing beyond our thoughts and sensations, how could we know anything about it? Empiricists believe that our senses provide us with our only source of factual knowledge, and many have thought that sensory evidence is not good enough for us to regard ourselves as having access to a world of the kind to which the realist is committed. And it seems strange (though not absurd, I think) to be in a position where you simultaneously say that a real world exists and also say we can never have any knowledge about it whatsoever.

In large part because of these debates about knowledge, various alternatives to realism were developed. "Idealism" is a term that can be used for several of them. *Subjective idealism*, as seen for example in the seventeenth-century empiricist George Berkeley, holds that the "physical" world is really comprised of a patterned flow of sensations, sent to us by God. Later, various philosophers argued for similar views (minus God) by arguing that all we could *mean* when we talk about "the real world" is some sort of regularity in our experiences. That position is sometimes called *phenomenalism*. If this is right, then when we seem to make claims about external objects, all we are talking about is patterns in our sensations. Another form of idealism, which came more out of opposition to empiricism, holds that the world amounts to more than a flow of sensations, but it nonetheless has a mental or spiritual character as a whole—I discussed these views back in chapter 2.

Part of what logical positivism wanted to do was claim that all these debates are meaningless. But the logical positivist theory of how language works, in at least some versions, seemed to bring the view rather close to phenomenalism—remember Schlick, quoted earlier in this book, saying that "what every scientist seeks, and seeks alone, are . . . the rules which govern the connection of experiences, and by which alone they can be predicted." Even after these views faded, empiricist views of the meaning of scientific language stayed around, and tended to encourage the idea that all we can ultimately talk about is our own experience, so science can never hope to describe the structure of a world beyond our senses.

Views like that did not do well as explanations of how science works. Sometimes people say that different views of the working of scientific language won't be reflected in the practice of science, but I think this is not right at all. As J. J. C. Smart, an early defender of scientific realism, argued, if a scientist was really a phenomenalist—did not merely say this, but acted with this commitment—that person would be much more conservative than most scientists, in a particular way. Suppose all you really want to do is get rules for the prediction of experience, and you encounter a problem—your predictions run into trouble. Why not just make a small patch, a small change to your theory, to deal with the problem? Why should you suspect that you have gotten something seriously wrong? If you merely patch up your theory, you will probably make it more complicated and inelegant than it was before, but this does not seem a very big deal if all you want to do is predict events. So what if your predictive machine is not as neat and simple as you'd like? Smart thought that a scientist who is a realist will treat problems differently—as raising the possibility, at least, that the theory's whole picture is wrong and should be reworked.

Empiricist views of scientific language, of the sort that cause problems with realism, do not fit well with how science works. But alternative views have been hard to develop. We are still waiting on a better theory of scientific language.

Not all empiricist critics of scientific realism have based their arguments in views of language. Bas van Fraassen, who has been an influential antirealist, confronts realism on the proper aims of science (1980). He suggests that *all we should ask* of theories is that they accurately describe the observable parts of the world. Theories that do this are "empirically adequate." An empirically adequate theory might also describe the hidden structure of reality, but whether or not it does so is of no interest to science. For van Fraassen, when a theory passes a lot of tests and becomes well established, the right attitude to have toward the theory is to "accept" it, in a special sense. (This is related to Laudan's sense of "acceptance," discussed in chapter 6, but not the same.) To accept a theory, for van Fraassen, is (1) to believe (provisionally) that the theory is empirically adequate, and (2) to use the concepts the theory provides when thinking about further problems and when trying to extend and refine the theory.

Regarding point 1, for a theory to be empirically adequate, it must describe *all* the observable phenomena that come within its domain, including those we have not yet investigated. Some of the familiar problems of induction and confirmation appear here. Regarding point 2, van Fraassen wants to recognize that scientists do come to "live inside" their theories; they make use of the theory's picture of the world when exploring new phenomena. But van Fraassen says a scientist can live inside a theory while remaining agnostic about whether the theory is true.

In discussions of scientific realism, the term "instrumentalism" is used to refer to a number of antirealist views. Sometimes it is used for traditional empiricist positions of the kind discussed above, but sometimes it is used in another way, one that I regard as more appropriate. Rather than saying that describing the real world is impossible, an instrumentalist will urge us not to worry about whether a theory is a true description of the world, or whether electrons "really exist." If we have a theory that gives us the right answers with respect to what we can observe, we might occasionally find ourselves wondering whether these right answers result from some deeper match between the theory and the world, but an instrumentalist thinks this question is not relevant to science.

Van Fraassen does not use the term "instrumentalist" to describe his view; he calls it *constructive empiricism*. The term "constructive" is used by so many people that it often seems to have no meaning at all, so I have reserved it for the views discussed below in section 10.5. I see van Fraassen's view as a kind of instrumentalism, but it does not matter much what we call it.

I think it is entirely possible for a scientist to aim only for empirical adequacy, in van Fraassen's sense. But he thinks that science should aim at no more than this. Why should we believe that? Why should science always stop at the observable? First, there is no sharp divide between the observable and unobservable parts of the world—unless you think that all you can observe is your own sensations. Some objects can be observed with the naked eye, like trees. Other things, like the smallest subatomic particles, can only have their presence inferred from their effects on the behavior of observable objects. But between the clear cases we have lots of unclear ones (Maxwell 1962). Is it observation if you use a telescope? How about a light microscope? An X-ray machine? An MRI scanner? An electron microscope?

Van Fraassen accepts that the distinction between the observable and the unobservable is vague, and he accepts that there is nothing unreal about the unobservable. He also accepts that we learn about this boundary from science itself. Still, he argues, science is only concerned with making true claims about the observable part of the world. But why should this be? Van Fraassen is saying that it's never reasonable for science to aim at describing the structure of the world beyond this particular boundary. Suppose we describe a slightly different boundary, one based on a concept a bit broader than observation. Let's say that something is *detectable* if it is observable or if its presence can be very reliably inferred from what is observable. As with van Fraassen's concept of observability, science itself tells us which things are detectable. In this sense, the chemical structures of various important molecules such as sugars and DNA are detectable although not observable. So why shouldn't science aim at giving us accurate representations of the detectable features of the world as well as the observable features?

Perhaps our beliefs about the detectable structures are not as reliable as our beliefs about the observable structures. If so, we need to be more cautious when we take theories to be telling us what the detectable structure of the world is like. But that is fine; we often need to be cautious.

What is so special about the "detectable"? Nothing, of course. We could define an even broader category of objects and structures, which includes the detectable things plus those that can have their presence inferred from observations with moderate reliability. Why should science stop before trying to work out what lies beyond this boundary? We might need to be even more careful with our beliefs about those features of the world, but that, again, is no problem.

You can see how the argument is going. There is no boundary that marks the distinction between features of the world that science can reasonably aim to tell us about and features that science cannot reasonably aim to tell us about. As we learn about the world, we also learn more and more about which parts of the world we can expect to have reliable information about. And there is no reason why science should not try to describe all the aspects of the world that we can hope to gain reliable information about. As we move from one area to another, we must often adjust our level of confidence. Sometimes, especially in areas such as theoretical physics, which are fraught with strange puzzles, we might have reason to adopt an instrumentalist view, or care only about empirical adequacy in van Fraassen's sense. But that does not apply across the board; we can often hope to do more.

10.5 *Metaphysical Constructivism*

I'll now look at a quite different way of being opposed to realism. I will use the term "metaphysical constructivism" for a family of views including those of Kuhn and Latour. These views hold that, in some sense, we have to regard the world as *created* or *constructed* by scientific theorizing. Kuhn expressed this claim by saying that when paradigms change, the world changes too. Latour expresses the view by saying that nature (the real world) is the product of the decisions made by scientists in the settlement of controversies. Nelson Goodman is another example; he argues that when we invent new languages and theories, we create new "worlds" as well (1978). For a metaphysical constructivist, it is not possible for a scientific theory to describe the world as it exists independent of thought, because reality itself is dependent on what people say and think.

These views are always hard to interpret, because they look so strange when we take them literally. How could we *make* a world (or the world) just by making up a theory? Maybe Kuhn, Latour, and Goodman are just using a metaphor of some kind? Perhaps. Kuhn sometimes expressed different views on the question, and he struggled to make his position clear. But when writers such as Goodman have been asked about this, they have generally insisted that their claims are not just metaphorical (1996, 145). They think there is something quite wrong with the realist picture. They accept that it's hard to describe a good alternative, but they think we should use the concept of "construction," or something like it, to express the relationship between theories and reality. The world as we know it and interact with it is partly (or entirely) a human construction. The way things are, or "the facts," is dependent on our beliefs, language, theories, or paradigms.

Some of these ideas can be seen as modified versions of the view of Immanuel Kant ([1781] 1998). Kant distinguished the "noumenal" world from the "phenomenal" world. The noumenal world is the world as it is in itself. This is a world that we are bound to believe in, but can never know anything about. The phenomenal world is the world as it appears to us. The phenomenal world is knowable, but it is partly our creation. It does not exist independently of the structure of our minds.

This kind of picture has often seemed appealing to philosophers who want to deny scientific realism, but do this in a moderate way. Hoyningen-Huene (1993) has argued that we should interpret Kuhn's views as similar to Kant's. In Michael Devitt's analysis of the realism

debates (1997), a wide range of philosophers are seen as either deliberately or inadvertently following the Kantian pattern. According to Devitt,
constructivist antirealism works by combining the Kantian picture with
a kind of relativism, with the idea that different people or communities
create different "phenomenal worlds" via the imposition of their different
concepts on experience. This relativist idea was not part of Kant's original
view; for Kant, all humans apply the same basic conceptual framework
and have no choice in the matter. The more recent, relativist views of this
kind give rise to the idea that each language community, each culture,
or perhaps even each person inhabits their own reality.

The Kantian picture is also seen as a way of holding on to the idea
that there is a real world *constraining* what we believe, but doing so in a
way that does not permit our knowing or representing this world. Though
that move can be tempting, it is hard to see how we make any progress
with these problems by adding an extra layer, "the phenomenal world"
or something similar, between us and the real world that constrains us.

The term "social constructivism" is often used for roughly the same
kind of view that I am calling metaphysical constructivism. But that term
is used for more moderate ideas as well. If someone argues that we make
or construct our theories, or our classifications of objects, that claim is
not opposed to scientific realism. We do indeed "construct" our ideas and
classifications; nature does not hand them to us on a platter. But realists
insist that beyond ideas and theories there is also the rest of the world,
much of which is not our product. In fields like sociology of science,
however, there is an unfortunate tradition of not explicitly distinguishing
between the construction of ideas and the construction of reality.

In history, philosophy, and sociology of science, I suggest that much
of the motivation for recent metaphysical constructivism, and also for
ways of writing that do not distinguish between the construction of ideas
and the construction of reality, lies in a determination to *oppose* a particular kind of view, and to emphatically signal one's opposition to it.
A lot of work in these fields has been organized around the desire to avoid
a picture in which reality determines what scientists and other people
think by stamping itself on the passive mind. The Bad View, the view
to avoid, holds that reality acts on scientific belief with "unmediated

compulsory force" (Shapin 1982, 163). That picture is to be avoided at all costs; it is often seen as not only false but culturally harmful, because it suggests a passive, uncreative view of human thought. Many traditional philosophical theories are seen as implicitly committed to this Bad View. This is one source for descriptions of logical positivism as reactionary, helpful to oppressors, and so on. A result is a tendency for people to try to go as far as possible away from the Bad View. This encourages people to assert simple *reversals* of the Bad View's relationship between mind and world. Thus we reach the idea that theories construct reality.

Some people explicitly embrace the idea of an "inversion" of the traditional picture (Woolgar 1988, 65), while others leave things more ambiguous. But there is little pressure within the field to discourage people from going too far in these statements. Indeed, those who express more moderate denials of the Bad View leave themselves vulnerable to criticism from within the field. The result is a literature in which one error— the view that reality stamps itself on the passive mind—is exchanged for another error, the view that thought or theory constructs reality.

Putting all this together, where does this leave me with scientific realism? Well, here is a summary of some things I believe. I accept common-sense realism. This could be undermined by science—I don't think that is likely, but it is possible in principle. To think that common-sense realism could be undermined by science is to take scientific theories as more than tools for prediction. More generally, I think that scientific theories, when things go well, can in principle and often do in fact tell us about how the world works and what it contains, including some of its deeply hidden structure. We have to be cautious about the messages of the parts of science that deal with the most inaccessible and esoteric entities, but that doesn't mean we can't hope to get things right even here. On the other hand, I don't accept the no-miracles argument for scientific realism discussed in section 10.3; that is too simplistic a view of the ways in which theories can be empirically successful. I think that phenomenalism, metaphysical constructivism, and other standard views that oppose realism are false. When those contrasts are in mind, it's natural to see myself as on the realist side, but as I said earlier in this chapter, once a person's responses to all these different issues are worked out, it's not very important to work out whether the view that results falls within the category of "scientific realism" or outside it.

10.6 *Underdetermination and Progress*

This chapter will end with two sections on particular topics in this area, one that was briefly discussed earlier and the other a little further away.

Above I said that scientific realism is often now seen (at least in large part) as a debate about whether we can be confident that we are getting things right—whether we should be optimistic or pessimistic about the aspirations of science to represent the world accurately. Some people argue that fundamental ideas have changed so often within science that we should always expect our current views to turn out to be wrong—just as many theories that people confidently believed in the past turned out to be wrong. Sometimes this argument is called the "pessimistic meta-induction." The prefix "meta" is misleading here, because the argument is not an induction about inductions; it's more like an induction about explanatory inferences. So let's call it "the pessimistic induction from the history of science." The pessimists give long lists of previously posited theoretical entities, like phlogiston and caloric, that we now think do not exist (Laudan 1981). Optimists reply with long lists of theoretical entities that once were questionable but which we now think definitely do exist, such as atoms, germs, and genes.

I said above that I think we should handle these questions in a field-by-field way; physics and biology have different sorts of histories and raise different questions. I'll now add some ideas on this topic.

Discussions of optimism and pessimism about science often take place against a backdrop provided by a general view, a claim called the "underdetermination of theory by evidence." In simple terms, this is the argument that for any collection of evidence we might have, there will always be more than one theory that can, in principle, accommodate that evidence. If so, this seems to show that our preference for any particular theory must always be influenced to some extent by factors other than evidence—by simplicity, elegance, or sheer familiarity (see Psillos 1999, chap. 8, for a review). This idea is closely related to some holistic ideas about testing discussed earlier in this book, and Quine, again, is often associated with this argument. In chapter 2 we looked at the claim that when things go wrong in our predictions, there are always several different ways to adjust our web of belief to make sense of what happened. Now we are looking at a slightly different idea: at any time, given the data we now have (whether these data include surprises or not), there will always be more than one theory we could choose that fits with our data.

A lot of discussion of this theme has looked at a sort of extreme case, at the idea that there can be different theories that are compatible with *all* possible evidence. Maybe there are real cases of this, but many of the examples discussed are somewhat artificial ones involving deceptive demons or very minor differences between two theories. A different "underdetermination" argument, one that I see as more relevant to science, has been developed by Kyle Stanford (2006). He argues that there is a situation that science finds itself in over and over again, one with troubling consequences. When we look back on previous episodes in the history of science, a common situation is that the evidence people had back then, which they used to make their choice of a theory, was also consistent with another theory that they had not ruled out, because they had not even thought of that theory. We can say this because we now *believe* the theory that no one had thought of back then. If this is how things appear when we look back, why shouldn't we expect that future scientists will look back on us the same way? That is, why shouldn't we expect future scientists to say that we, today, should not have been so confident in our choices, because they (scientists in the future) will then be using a theory that we currently can't imagine? Stanford calls this the argument from "recurrent transient underdetermination." The argument is a sort of combination of the pessimistic induction from the history of science and the idea of underdetermination.

This will not be much of a problem if the usual case is one where the future theory is an improvement on earlier ideas, of a kind that shows the earlier ideas to have been on the right track. Most people working now would not mind being told later that their ideas were on the right track but needed improvement. The argument is troubling if the typical case is one where new theories tend to flatly reject the main ideas in old ones. That question can only be resolved by looking closely at the history of science, in different areas. We also need to keep in mind the possibility that a theory can appear, in retrospect, to have been quite misguided about what sorts of things the world contains but nonetheless have some structural features that we can recognize as being on the right track. A theory can get us close to some crucial relationships while being wrong in other ways (Worrall 1989).

There is also another way in which these underdetermination claims can be misleading. Stanford argues, as discussed above, that at each moment our data will be compatible with various different theories, including some we have not yet thought of. That is how things usually look in retrospect. But it's also compatible with this historical picture that whenever people find there are alternative theories that are compatible with all the current data, they go out and look for more data, and make a new choice between the theories. This will not be the end of the process, because the new collection of data will also be compatible with more than one theory. But then we can get even *more* data and try to discriminate between *those* alternatives. In this sequence of events, there is always underdetermination—we never get down to one theory. But there might seem to be a kind of progress, too. How significant this progress is might depend a lot on the relations between theories that we choose at different times—whether theories build on their predecessors, or just replace them. This affects whether we are really heading somewhere, or just traveling.

We could go further down this path, but I think there are some problems with the whole way the situation is being set up, and soon it will be time to correct them. That will happen in chapter 12, when we come to modern theories of evidence. Once we've examined those views, a lot of things in this section will look different.

10.7 *Natural Kinds*

To finish this chapter I'll look at a very different theme. In some discussions of realism, especially as it relates to science, people say that a
realist position includes the view that the world is neatly divided up into
"natural kinds," into categories (like *electron, gene,* and so on) that our
scientific theories give labels to. Realism is said to include the idea that
nature has definite "joints" or "seams," and an ideal scientific language
would give us a perfect partitioning of the world—a categorization that
God would agree with, if God wanted to do science. Sometimes the term
"metaphysical realism" is used for this sort of position.

There is something right here. It would be an extremely minimal realism that just said there is *a* world we live in and deal with, perhaps one
with no definite structure. But there are reasonable versions of realism
that do not say that the world is prepackaged into neat kinds, or anything
like that. And however this topic relates to realism, the problem of kinds
and categories is interesting in its own right.

One of the main things language does is provide us with groupings of
objects—words such as "human," "enzyme," and "automobile" do this.
The term "nominalism" is sometimes used in philosophy for the idea
that groupings are entirely imposed by us; any collection of things can be
given a name, and the world does not compel us to categorize things in
one way rather than another. An alternative view on this question is that
some groupings are more natural than others—more in accordance with
the way the world is. A view of this sort is often seen as both reasonable
in its own right and potentially useful in solving other philosophical
problems. Back in chapter 3, I mentioned attempts to solve Goodman's
new riddle of induction with the idea that the word "green" picks out a
natural kind while the word "grue" does not. More standard examples
of natural kinds are the chemical elements—*hydrogen, helium,* and so
on through the periodic table. (I use quotation marks when I am talking
about words, like "hydrogen," and italics when I am talking about kinds
or collections of things in the world, like *hydrogen.*) If one is looking for
natural kinds, the chemical elements do seem to be good examples. To
claim that they are natural kinds is not just to say that hydrogen atoms

are real, but that the kind *hydrogen* is real in a mind-independent way also—our word "hydrogen" picks out a collection of things that are unified by real similarity. In many sciences, the question of whether standard terms pick out real or natural kinds is the topic of ongoing debate, even if the philosophical terminology "natural kind" is not used. For example, are the mental disorders categorized in psychiatric reference books such as the *Diagnostic and Statistical Manual* (also known as the "DSM") natural kinds, or do we tend to apply labels like "schizophrenia" to a range of cases that have no real underlying similarity?

The example of schizophrenia also points us to another feature of the situation. In cases like this that involve the study of human beings, there are two varieties of dependence that a kind or grouping might be said to have upon us. First, there is the question of whether the category is arbitrary from a medical and biological point of view. Second, there is a separate question of what the effects are, on that collection of people, of having that categorization put in place. If a collection of people are categorized and described in a particular way, whether the grouping was initially arbitrary or not, this can have consequences of its own. These consequences can be mediated by both the treatment of those people by others (by medical workers, in the case of schizophrenia) and by those people's own self-conception.

Ian Hacking discusses this phenomenon, especially in relation to mental illnesses, in his book *The Social Construction of What?* (1999). Hacking distinguishes what he calls "interactive" kinds from "indifferent" kinds. *Schizophrenia* is an interactive kind; when the term "schizophrenic" is introduced and applied to people, it can lead to new behaviors in those people as they respond to this categorization. *Hydrogen* is an indifferent kind; it does not respond to our categorization in these ways. Hacking marks out interactive kinds in a narrow way; the members of the kind have to know and care about their classification. There is also a broader phenomenon here, in which the path through the world of a collection of objects (crop plants, for example, or an endangered species) is materially affected by our forming a new classification of them, and hence behaving differently toward them, even though the objects do not know it.

A problem with many discussions in this general area is the fact that there appears to be a lot of real *structure* in the world that is not organized into kinds. To pick a very simple example, differences in height between people are real (they do not depend on what we say and think), but tall people are not a natural kind. There is no border between the tall and the non-tall; there are just lots of gradations in height. Words like "tall," as opposed to "taller than," are almost always vague and indefinite in their application. The structure that exists in the realm of human height lends itself to mathematical description, and that, of course, is the usual way of describing height—in feet or centimeters. Part of the power of mathematics lies in its ability to represent structure of this sort.

Many of these questions about kinds, construction, interaction, and mind independence are brought together in a case that has been discussed extensively over recent years. This is the status of human racial categories, such as *black*, *white*, and so on. The discussion has gone through several phases. Back in the latter part of the twentieth century, attention to emerging knowledge in human genetics and an awareness of the artificiality of traditional racial categories led to a number of people arguing that races, as traditionally understood, do not exist at all (Lewontin 1972; Appiah 1994). The whole idea of races as *kinds* of people was seen as a mistake. This is sometimes called "eliminativism" about race— the traditional categories ought to be eliminated. More recently, this view has been seen as failing to recognize a kind of reality that races have as a consequence of human beliefs and institutions. "Constructivist" views of race hold that although human races are not biologically natural, they are nonetheless real, as a consequence of human politics, history, and attitudes. Being put into a racial category has consequences that are not in any sense illusory, though they do depend on human attitudes.

The views distinguished above are often presented as options that one should choose between—the literature asks us to choose between being a realist, a constructivist, or an eliminativist about race. But it seems better to recognize a number of different "kinds of kinds" here, with distinct roles. Michael Hardimon (2017) recognizes no fewer than four. In his framework, *racialist races* are groups of humans that are supposed to have different underlying biological natures that are associated with different abilities and moral characters. These racialist races, for Hardimon, do not exist; he is an eliminativist about racialist races. There is also a *minimalist* concept of race, in which people of a particular race merely have some recognizable physical differences from others, indicating a particular geographical ancestry. Hardimon thinks that minimalist races are real, and also thinks this is the everyday concept of race. So he is not an eliminativist about these. *Populationist races* are collections of people with genetic similarities that reflect a particular ancestry; this is a biological analog of minimalist race. He thinks these races are real also. Lastly, there are what he calls *socialraces*. A socialrace is a collection of people who have been *taken* to comprise a racialist race (the first of his concepts above). Being a member of one of these groups can have many consequences, even though racialist races do not exist.

In a simpler framework, Adam Hochman (2017) uses the term "racialized group" for a group of people who have been treated as forming a race in the traditional and discredited conception. Those things—racialized groups—are real and consequential, even though the assumptions that formed the original basis for the categorizations are erroneous. These groupings are not "natural" kinds in a biological sense, but they are "natural" in not being merely arbitrary. The ways people are treated on the basis of racial assumptions and categorizations are genuine aspects of human social life and its history.

Much thinking about human races also tends to look for definite boundaries and kinds, even when there are gradations and mixtures. This is certainly true of the discredited racialist races in Hardimon's sense, but not only of those. In this respect, even Hardimon's minimalist races might in some cases be questionable.

The example of race is especially tangled, but it illustrates something general. In many contexts, a distinction between "natural" kinds and arbitrary or non-natural groupings is insufficient to capture what is going on.

Further Readings and Notes

Central works in the mid-twentieth-century resurgence of scientific realism include J. J. C. Smart's *Philosophy and Scientific Realism* (1963) and various papers collected in Hilary Putnam's *Mind, Language, and Reality*

(1975b). Leplin (1984) is a good collection on the debate as it developed in those years. In that book, Boyd (1984) gives an influential defense of scientific realism in the no-miracles tradition.

Devitt's *Realism and Truth* (1997) defends both common-sense and scientific realism. Psillos (1999) gives a very detailed treatment. Hacking (1983) and Fine (1984) are influential works on scientific realism that defend rather different views from those discussed here. Hacking focuses on the role of experimental intervention, while Fine tries to defuse the debate by criticizing both realist and antirealist commentaries on science.

On the puzzles surrounding quantum mechanics, Becker (2018) is helpful, along with the Maudlin book cited earlier (2019). I said that my understanding is that interpretations of quantum mechanics according to which the state of the world is dependent on the perceptions of scientists performing measurements are fading in influence, though as I was writing this chapter, a new paper defending this message appeared: Proietti et al. (2019). Schrödinger and Wigner took this view seriously. Einstein accused Bohr of a view like this, but Bohr apparently said Einstein had misunderstood him; see Schrödinger (1958) and Faye (2019). Here is Wigner:

> It was not possible to formulate the laws of quantum mechanics in
> a fully consistent way without reference to the consciousness. . . .
> It may be premature to believe that the present philosophy of
> quantum mechanics will remain a permanent feature of future
> physical theories; it will remain remarkable, in whatever way our
> future concepts may develop, that the very study of the external
> world led to the conclusion that the content of the consciousness is
> an ultimate reality. ([1961] 1995, 172)

One of the most popular recent approaches to the problem of natural kinds is the "homeostatic property cluster" view developed by Boyd (1991) and others. I don't think this approach helps very much, because the idea of "homeostasis" is merely a metaphor in most cases outside of some particular parts of biology. On the issue of gradients as opposed to kinds in biology, see Gannett (2003). For additional discussions of race and racial kinds, see Spencer (2018a and 2018b), and Pigliucci and Kaplan (2003). For other discussions of properties and kinds, and their relevance to induction, see Armstrong (1983 and 1989), Dupré (1993), and Kornblith (1993).

Chapter 11

Explanation, Laws, and Causes

11.1 *Knowing Why*

What does science do for us? A simple answer is that science tries to tell us what the world contains and what goes on in it. Perhaps the goal is narrower; some have argued that science only tries to describe the observable part of the world. But whatever you think about this question, a further issue arises. It is common to think that science tells us *why* things happen; we learn from science not just *what* goes on, but why it does. Science apparently seeks to explain as well as describe. What is it for a scientific theory to explain something? In what sense does science give us an understanding of phenomena, as opposed to mere descriptions of what happens?

The idea that science aims at explanations of why things happen has sometimes aroused suspicion in philosophers, and also in scientists themselves. Such distrust is reasonably common within empiricist views. Empiricists have quite often seen science, most fundamentally, as a system of rules for predicting events. When explanation is put forward as an extra goal for scientific theories, some empiricists get nervous. Or perhaps, as other empiricists say, explanation and prediction are so closely related that there is no problem here.

There is a complicated relationship between this problem about explanation and the problem of analyzing evidence. The hope has often been to treat these problems separately. Understanding evidence is problem 1; this is the problem of analyzing when you have good reason to believe that a scientific theory is true. Understanding explanation is problem 2; here we assume that we have already chosen our scientific theories, at least for now. We want to work out when and how our theories provide explanations. A distinction of this kind can be made, but there is a close connection between the issues. The solution to problem 2 may affect how we solve problem 1. Theories are often preferred by scientists because they seem to yield good explanations of puzzling phenomena. In chapter 3, *explanatory inference* was defined as inference from some data to a hypothesis about a structure or process that would explain the data. This seems much more common in science than purely inductive inference (inference from particular cases to generalizations). That fact suggests that there is a close relation between the problem of analyzing explanation and the problem of analyzing evidence.

This chapter will mostly be about explanation itself, trying to set aside questions about the relation between explanation and evidence. Evidence will return to center stage in chapter 12. Two other topics that are closely linked to explanation are causation and laws of nature. Perhaps to explain something is to describe its causes, or the laws behind it. But what are causes and laws? Those questions will be discussed in the last part of the chapter.

11.2 *The Rise and Fall of the Covering Law Theory of Explanation*

Empiricist philosophers, I said above, have sometimes been distrustful of the idea that science explains things. Logical positivism is an example. The idea of explanation was sometimes associated by the positivists with the idea of achieving deep metaphysical insight into the world—an idea they would have nothing to do with. But most of these philosophers did make peace with the idea that science explains. They did this by construing "explanation" in a low-key way that fitted into their empiricist picture.

The result was the *covering law theory* of explanation. This was the dominant philosophical theory about scientific explanation for a good part of the twentieth century. It outlived the logical empiricist view that gave rise to it. The view is now dead, but its rise and fall are instructive.

The covering law theory was first developed in detail by Carl Hempel and Paul Oppenheim (1948). Let us begin with some terminology. In talking about how explanation works, the *explanandum* is whatever is being explained. The *explanans* is the thing that is doing the explaining. If we ask "why *X*?," then *X* is the explanandum. If we answer "because *Y*," then *Y* is the explanans.

The basic ideas of the covering law theory are simple. First, to explain something is to show how to derive it in a logical argument. The explanandum will be the conclusion of the argument, and the premises are the explanans. A good explanation must first of all be a good logical argument, but in addition, the premises must contain at least one statement of a *law of nature*. The law must make a real contribution to the argument; it cannot be something merely tacked on. Of course, for an explanation to be a good one in the fullest sense, the premises must also all be true. But the first task here is to describe what sorts of statements *would* give us a good explanation of a phenomenon, if the statements were true.

Some explanations (both in science and everyday life) are of particular events, while others are directed at general phenomena or regularities. For example, we might try to explain the particular fact that the U.S. stock market crashed in 1929 in terms of economic laws operating against the background conditions of the day. And we can also explain patterns; Newton is often seen as giving an explanation of Kepler's laws of planetary motion in terms of more basic laws of mechanics in combination with assumptions about the layout of the solar system. In both cases, the covering law theory sees these explanations as expressible in terms of arguments from premises to conclusions. Some of the arguments that express explanations will be deductively valid, but this is not required in all cases. The covering law theory was intended to allow that some good explanations could be expressed as nondeductive arguments ("inductive" arguments, in the logical empiricists' broad sense of the term). If we can take a particular phenomenon and embed it within an argument in which the premises include a law and bestow high probability on the conclusion, this yields a good explanation of the phenomenon.

Many problems in the details were encountered in attempts to formulate the covering law theory precisely (Salmon 1989). The problems were more difficult in the case of nondeductive arguments, and also in the case of explaining patterns rather than particular events. I won't worry about the technicalities here. The basic idea of the covering law theory is simple and clear: to explain something is to show how to derive it in a logical argument that makes use of a law in the premises. To explain something is to show that it is to be expected—to show that it is not surprising, given our knowledge of the laws of nature.

For the covering law theory, there is not much difference between explanation and prediction. To predict something, we use an argument to try to show that it is to be expected, though we don't know for sure yet whether it is going to happen. When we explain something, we know that it has happened, and we show that it could have been predicted using an argument containing a law. You might be wondering at this point what a "law of nature" is supposed to be. This was a troubling topic for logical empiricism, and has continued to be troubling for everyone else. But a law of nature, for a logical empiricist, was not supposed to be something very grandiose. It was supposed to be a kind of basic regularity, a pattern, in the flow of events. (I return to this question in section 11.5.)

Though I use the phrase "covering law theory" here, another name for the theory is the "D-N" theory, or model, of explanation. "D-N" stands for "deductive-nomological," where "nomological" is from the Greek word for law, *nomos*. The term "D-N" can be confusing because the argument in a good explanation need not be deductive. So "D-N" really only refers to some covering law explanations: the deductive ones.

That concludes my sketch of the covering law theory. I now move on to what is wrong with it. This is a case where we have something close to a knockdown argument. Although there are many famous problems with the covering law theory, the most convincing one is usually called the *asymmetry problem*. And the most famous illustration of the asymmetry problem is the case of the flagpole and the shadow.

Suppose we have a flagpole casting a shadow on a sunny day. Someone asks: why is the shadow *X* meters long? According to the covering law theory, we can give a good explanation of the shadow by deducing the length of the shadow from the height of the flagpole, the position of the sun, the laws of optics, and some trigonometry. We can show why that length of shadow was to be expected, given the laws and the circumstances. The argument can even be made deductively valid. So far, so good. The problem is that we can run just as good an argument in another direction. Just as we can deduce the length of the shadow from the height of the pole (plus optics and trigonometry), we can deduce the height of the pole from the length of the shadow (and the same laws). An equally good *argument*, logically speaking, can be run in both directions; either can give information about the other. But it seems that we cannot run an equally good *explanation* in both directions, though the covering law theory says we can. It is fine to explain the length of the shadow in terms of the flagpole and the sun, but it is not fine to explain the length of the flagpole in terms of the shadow and the sun. (At least, it is not fine unless this is a very unusual flagpole—perhaps one that is designed to regulate its own length in a way that maintains a particular shadow.)

What we find here is that explanations have a kind of directionality. Some arguments (though not all) can be reversed and remain good as arguments. But explanations cannot be reversed in this way (except in special cases). So not all good arguments that contain laws are good explanations. This objection to the covering law theory was given (using a slightly different example) by Sylvain Bromberger (1966).

Once this point is seen, it becomes obvious and devastating. The covering law theory sees explanation as very similar to prediction; the only difference is what you know and what you don't know. But this is a mistake. Consider the concept of a *symptom*. Symptoms can be used to predict, but they cannot be used to explain. Yet a symptom can often be used in a good logical argument, along with a law, to show that something is to be expected. If you know that only disease D produces symptom S, then you can make inferences from S to D. You might in some cases be able to make predictions from D to S, too. But you cannot explain a disease in terms of a symptom. Explanation only runs one way, from D to S, no matter how many different kinds of inferences can be made in other directions. And, further, it seems that good explanations of S can be given in terms of D even if S is *not* a very reliable symptom of D, even if S is not always to be expected when someone has D. This is a separate problem for the covering law theory, often discussed using the example of some unreliable but devastating symptoms of the disease syphilis.

In some of these cases, the covering law theory can engage in footwork to evade the problem. But other cases, including the original flagpole case, seem immune to footwork. Hempel's own attitude to the issue was puzzling. He actually anticipated the problem, but dismissed it (Hempel

1965, 352–54). His strategy was to accept that if his theory allowed explanations to run in two directions in cases where it seems that explanation only runs in one direction, then both directions must be OK. In some actual scientific cases this reply seems reasonable; there are cases in physics where it is hard to tell which direction(s) the explanation(s) are running in. But in other cases the direction seems completely clear. In the case of the flagpole and the shadow, this reply seems hopeless. Hempel might reply that this is not a case relevant to science. But the relation between diseases and symptoms is relevant to science, and as we saw above, this can be used to make the point too. There are other good arguments against the covering law theory (Salmon 1989), but the asymmetry problem is the most important. It also seems to be pointing us toward a better account of explanation.

11.3 *Causation and Unification*

What is it about the flagpole's height that makes it a good explanation for the length of the shadow, and not vice versa? The answer seems straightforward. The shadow is caused by the interaction between sunlight and the flagpole. That is the direction of causation in this case, and that is the direction of explanation, too. We seem to get an immediate suggestion from the flagpole case for how to build a better theory: to explain something is to describe what caused it. Why did the dinosaurs become extinct 65 million years ago? Here again, our request for an explanation seems to amount to a request for information about what caused the extinction.

   Although that conclusion seems rather compelling, it has not been universally accepted, and it raises further problems. What is causation? It seems to make a lot of sense to use the idea of causation to resolve the flagpole case, but for many philosophers, causation is a suspicious metaphysical concept that we do best to avoid when trying to understand science. Suspicion is directed especially at the idea of causation as a sort of hidden connection between things, unobservable but essential to the operation of the universe. Empiricists have often tried to understand science without supposing that science concerns itself with alleged connections of that kind. The rise of scientific realism in the latter part of the twentieth century led to some easing of this anxiety. But many philosophers would be pleased to see an adequate account of science that did not get entangled with issues about causation.

Despite this unease, as people abandoned the covering law theory, the main proposal about explanation discussed, in different ways, was the idea that explaining something is giving information about how it was caused (Salmon 1984; Suppes 1984). It might seem initially that this view of explanation is most directly applied to explanations of particular events (such as the extinction of the dinosaurs), but it can also be applied to the explanation of patterns. We can ask, why is inbreeding associated with an increase in birth defects? The explanation will describe a kind of causal process that is involved in producing the phenomenon (a process involving an increased chance that two copies of a harmful recessive gene will be brought together in a single individual).

Even if we knew what causation is, this would not settle all the questions about causation and explanation. We also need to know what kind of information about causes makes for a good explanation. An approach to this question that makes good sense was developed by Peter Railton (1981). We first imagine an idealized "complete" explanation that contains everything in the causal history of the event to be explained, specified in total detail. No one ever wants to be told the complete explanation for a phenomenon, and we never know these complete explanations. That's fine; instead, in any context of discussion in which a request for an explanation is made, some piece or pieces of the complete explanation will be relevant. We are often able to know, and describe, these relevant pieces of a total causal structure. To give a good explanation in actual practice, all that is required is a description of these relevant pieces of the whole.

In the years after the demise of the covering law theory, much work focused on causation. But an alternative project developed, especially among those wary of the concept of causation. This was the idea that explanation should be analyzed in terms of *unification*. This approach was developed especially by Michael Friedman (1974) and Philip Kitcher

(1981 and 1989), though as Kitcher emphasized, it was actually present all along in logical empiricism. The view that explanation is unification was a sort of "unofficial" theory of explanation within much logical empiricism, in contrast to the official covering law theory (see, for example, Feigl 1943). The unofficial theory is a good deal better than the official one. Often the two approaches were mixed in together; to show the connection between particular events and a general law is, after all, to achieve a kind of unification. But why not develop a theory of unification in science that is not tied to the idea of deriving phenomena from laws? The *unificationist* theory holds that explanation in science is a matter of connecting a diverse set of facts by subsuming them under a set of basic patterns and principles. Science constantly strives to reduce the number of things that we must accept as fundamental. We try to develop general *explanatory schemes* that can be applied as widely as possible.

The search for unity is certainly an important feature of science, and what produces an "aha!" reaction in science is often the realization that some odd-looking phenomenon is really a case of something more general. As Kitcher argued, some very famous theories—Darwin's theory of evolution and Newton's later work on the nature of matter—were compelling to scientists in their early stages despite not making many new predictions, because they promised to explain so much. And this "explanatory promise" seems to have been the ability of those theories to unify a great range of phenomena with a few general principles. In Kitcher's case, another reason for developing the unificationist theory was a distrust of the idea of causation. This led Kitcher to try, for some years, to develop a theory of explanation entirely in terms of unification. What about the flagpole and the shadow, and the asymmetry in which can explain which? Kitcher (1989) argued that we do tend to describe this asymmetry in causal terms, but this causal talk is really a loose summary of more basic asymmetries that involve unification.

This gives us two main proposals to replace the covering law theory: the causal theory and the unification theory. These have often been treated as competitors: does causation win, or does unification win? But I think this is a mistake; we do not have to choose. Again, beware the dubious allure of simplicity in philosophical theories. Much of the time, to explain something is to describe the causal mechanisms behind it or the causal history leading up to it. That is true much of the time, but there is no need to hold that it is true all of the time. In some cases there can be pretty clear explanatory relations between patterns or principles, even when it is hard to apply causal language to the situation. Often this seems to involve unification. Nothing stops us from holding that a variety of different relations can be explanatory. Some proponents of causal and unificationist theories eventually came to accept a view of this kind (Salmon 1998; Kitcher 2001).

The view that then emerges is a kind of *pluralism* about explanation in the philosophy of science. That is a step in the right direction, but I think the familiar forms of pluralism here are not quite right. I think that just as it was a mistake to think explanation is the kind of thing that requires analysis in terms of a single special relation, it is a mistake to think there are two or three such relations that can be freely chosen between. An alternative is to say that the idea of explanation operates differently within different parts of science, and differently within the same part of science at different times. The word "explanation" is used in science for something that is sought by the development of theories, but what exactly is being sought is not constant in all of science. We cannot get the right analysis by claiming that within science as a whole, a good explanation is something that satisfies *either* the causal test *or* the unification test (etc.). That form of pluralism leaves out the way that different scientific fields will establish their own criteria for what will pass as a good explanation. The standards for a good explanation in field *A* need not suffice in field *B*. If an "ism" is required, the right analysis of explanation is a kind of *contextualism*—a view that treats the standards for good explanation as partially dependent on the scientific context.

Kuhn argued some years ago for a view of this kind (1977a). In a paper about the history of physics, he claimed that different theories (or paradigms) tend to bring with them their own standards for what counts as a good explanation. He argued, further, that standards about whether a relation counts as "causal" also depend on paradigms. The concepts of explanation and causation are, to some extent at least, internal to different scientific fields and historical periods.

In the case of causation, a philosopher might reply to Kuhn that just because different people have thought differently about causation does not mean that there is no fact of the matter. Perhaps this is right. But in the case of explanation, I think this reply has little force. If two scientific fields single out different relations and call them "explanation," there need be no factual error that one or the other is making.

To support his claims, Kuhn focused on Newton's theory of gravity. Does Newton's theory explain the falling of objects, given that Newton's treatment of gravity gave no intuitive mechanism but only a mathematical relationship? Some people answered no, but over time it became part of Newtonianism that the right kind of mathematical law does count as an explanation. A similar attitude was developed about Maxwell's equations describing electromagnetism. Kuhn's view is that the idea of explanation will evolve as our ideas about science and about the universe change. So although the covering law theory definitely fails as a general account of explanation in science, it would be a mistake to conclude that *no* explanations have the form described by the covering law theory. Some explanations are at least close to what Hempel had in mind; the mistake is to apply that model to all cases.

This proposal need not lead to the radical idea that anything can count as an explanation. Scientific traditions will generally have good reasons for their treatment of the idea of explanation, and views about explanation will depend on views about what the world contains. Some possible concepts of explanation will embed factual errors. To use a simple example, if someone claims that good explanations are always based on a concept of God's will, but there is no God, then that conception of explanation will be mistaken because of a factual error. Some philosophers might make the same argument about concepts of explanation that use the idea of causation—they might argue that the traditional idea of causal connection is a piece of mistaken metaphysics.

It is worthwhile to pay attention to the actual use of the term "explain" in science. Here we find a lot of diversity. In some fields, there are technical senses of the word, even mathematical measurements of "the amount of variation explained" by a given factor. In other fields, nothing like a technical standard applies. The word "explain" also has a kind of rhetorical use. Someone might say: "Your theory does accommodate this result, but it does not really explain it." This might mean, "Your theory can only be used to derive this result in a very unnatural-looking way." (Often unification seems important in cases like this.) At other times, the word "explain" is used in a much more low-key manner. For example, a lot of science is aimed at describing how things work. How does photosynthesis work? How does the replication of DNA work? Descriptions of phenomena of this kind will often be referred to as explanations, but this does not mean that something is going on beyond a description of mechanisms and processes.

I will take a moment to compare my view with another unorthodox position in this area, that of van Fraassen (1980). He denies, as I do, that explanation is some single, special relation common to all of science. He has developed a "pragmatic" account of explanation, where what counts as an explanation varies according to context. But this is different from my view. Van Fraassen wants to deny that explanation is something *inside* science at all; he denies that scientific reasoning includes the assessment of the explanatory power of theories. Instead, explanation is something that people do when they take scientific theories and use them to answer questions that are external to scientific discussion itself. In contrast, the view I am defending is one in which explanation is thoroughly internal to science, but *variously so*. Assessments of what explains what are an important part of scientific reasoning, but different fields may use somewhat different concepts and standards of explanation.

Before leaving this topic, I should add another comment about *explanatory inference*. Back in chapter 3, I used this term for inference from a set of data to a hypothesis about a structure or process that would explain the data. The term "explanatory inference" suggests that there is one kind of relationship between data and hypothesis—*the* explanation relationship—that is involved in explanatory inference. Many philosophers do accept this. The term "inference to the best explanation" is, in fact, a more common name for what I call explanatory inference, and that term suggests that there is a single measure of "explanatory goodness" involved; the option that is best as an explanation is the one most likely to be true. But I think this is the wrong way to think about scientific reasoning (and that is why I have avoided the term "inference to the best explanation"). I use "explanatory inference" in a broader way that does not suppose that there is a single measure of explanatory goodness, which applies to all of science. Rather, explanatory inference is a matter of devising and comparing hypotheses about hidden structures that might be responsible for data. "Explanation" is seen here as something pretty diverse.

To sum up: the covering law theory is dead, and its death was informative. In its wake, we need not look for a new theory of some single relation or pattern that is involved in all scientific explanation. Very often, causation is involved. The same goes for unification and for deriving phenomena from laws. But different fields have different concepts and standards of explanation.

11.4 *Laws, Causes, and Interventions*

The covering law theory of explanation made use of the idea of a law of nature. One of the theories that replaced it made use of the idea of causation. But what are laws of nature? What is causation?

In both cases, we have concepts that seem aimed at picking out a special kind of *connection* between things in the world. Causation is sometimes called, half jokingly, "the cement of the universe" (Mackie 1980; the phrase was first used by David Hume [(1740) 1978]). Many philosophers have been somewhat skeptical about these concepts. Some have tried to work without them, while others have not rejected them but tried to reconstrue them in a very low-key way ("Yes, there is causation, but it is no more than this . . ."). In particular, philosophers have tried to analyze both laws and causation in terms of patterns in the *arrangement* of things, rather than some extra connection *between* things. Sometimes this project is referred to as "Humeanism," as Hume was the first philosopher to develop a really focused suspicion about concepts of connection between events in nature (Lewis 1986b). Present-day Humeans do not have the same kind of empiricism as Hume, but they do want to avoid believing in any sort of unobservable cement connecting the universe together.

A philosopher following this approach will try to construe laws of nature as no more than regularities, or basic patterns, in the arrangement of events. To treat laws of nature in this way is to leave behind one of the familiar connotations of the term "law." Usually, we see laws as directing, guiding, or governing in some way. In this spirit, laws of nature can be seen as governing the flow of events and hence responsible for the regular patterns that we see, rather than being identical with those patterns (Dretske 1977; Armstrong 1983). Those following the Humean approach see a "governing" conception of laws as a seduction that must be avoided. The logical empiricist attitude toward laws of nature was Humean in this sense.

Whatever laws might be, for some years philosophers tended to discuss their role in a way that took it for granted that laws are central to science. In 1983, Nancy Cartwright delivered a wake-up call to the field with a book called *How the Laws of Physics Lie*, in which she argued that what people call "laws of physics" do not usually describe the behavior of real systems at all, but only the behavior of highly idealized or fictional systems. Another important change that resulted from a closer look at actual science is that philosophers are no longer obsessed with natural laws as *the* goal of scientific theorizing. For many years philosophers searched fields such as biology for statements of laws of nature. Philosophers thought that any genuine science had to contain hypothesized laws, and had to organize its main ideas in terms of laws. In fact, most biology has little use for the concept of a law of nature, but that does not make it any less scientific.

The topic of causation has generated debates that are sometimes similar to debates about laws. On one side, we have those who see causation as no more than a kind of regular pattern in the flow of events. On the other side are those who see causation as a connection between events— between the ball hitting the window and the glass breaking—that is *responsible* for the patterns. That is similar to the main debate about laws, and sometimes people have seen laws and causes as just two sides

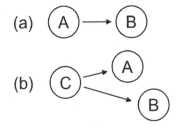

Figure 11.1, a and b. Two causal possibilities

of the same coin. A difference is that in the case of causal relations, it's possible to argue that there are indeed hidden connections between things, but not so hidden that they can't be studied directly by physics (Dowe 1992).

Over recent decades, another approach to causation has been developed in detail and has yielded considerable progress, fusing various parts of science, philosophy, and statistics. This is the *interventionist* approach (also sometimes called *manipulationist*).

Imagine you are dealing with a system of some sort in which you know roughly what factors are present, and the question is what the causal relations are between them. Suppose, in particular, that you want to know whether *A* is a cause of *B*. When I say "*A*" here, I mean to refer not to just one event, or one kind of event, but what can be called a *variable*. An example of a variable is *the weather,* which might be hot or cold on a particular day. A variable might also be a behavior, such as smoking or not smoking, or a disease, which might be present or absent. You want to know whether the situation can be pictured as it is in figure 11.1a, in which arrows indicate a causal relationship between *A* and *B*. A problem you face is that the situation might instead follow figure 11.1b. Here, there is no arrow from *A* to *B*, but another variable, *C*, affects both *A* and *B*. (It's also possible that *B* is affected by both, but I am dealing with the simplest cases here.)

Imagine being able to reach in and change the state of just one variable or one "node" in the network. When you reach in and change the state of *A*, if you find that *B* changes (or becomes more likely), you have found that *A* is a cause of *B*. Causal analysis, on this view, is analysis of what follows from interventions, whether those interventions are real or imagined.

Why do we have to imagine an intervention, a reaching in? Why don't we just imagine a change *happening* to one part of the network, *A*, and ask whether this change to *A* is followed by changes to *B*? Why not say that when a change to *A* is followed by a change to *B*, *A* is a cause of *B*? This is not enough, because when *A* changes and *B* also changes, that might be because some other factor *C* affects both *A* and *B*, as in the diagram, or perhaps there is some other pathway that brings about the association. In the interventionist way of approaching these questions, a central distinction made is between the case where *A* changes and *B* changes because *A* changed, and the case where *A* changed and *B* changed for some other reason. This is a development of the well-known slogan that "correlation is not causation"; the interventionist framework is an attempt to deal rigorously with those relationships. An intervention is supposed to be a kind of "surgical" change—we change *A* alone and see what ensues.

As a way of dealing with causation, this approach has been fruitful in a number of ways. It appears that very young children apply something like these principles when they explore causal relationships (Gopnik and Schulz 2007). It has led to new methods within science itself (Pearl 2000). It also has interesting connections to a question I raised back at the end of chapter 2: what is the relationship in science between observational data, in a broad sense, and data we get from experiments? We saw Feynman saying that experiment is essential to science, a view that presents a number of problems. (What about astronomy? And *why* should experiments be special?) From an interventionist point of view, experiments that manipulate a situation are special because interventions can unearth causal relationships in a way that non-experimental data often cannot.

When you do an experiment, you bring about localized change to some part of the world, and see what happens next. You do not merely look on when *A* changes and see what goes along with it. Experiment is not absolutely essential to learning about causation; in some cases you might have background knowledge that enables you to discern the relationships without actual intervention—you might know there is no additional factor *C* that could lie behind both *A* and *B*. If you have a lot of information about probabilities of a certain kind, information that does not require interventions, you can work out the location and direction of causal arrows in a graph, even a complex one. You would need to know not just the overall probabilities of things happening, but the conditional probabilities—such as the probability of *A given C*, and so on. (We will look at probability in the next chapter.) There can sometimes also be "natural experiments," when a change that can be seen to be localized in the right way just happens to come about. But just as interventions are often the best way of working out how the factors in figure 11.1 are set up, deliberate experiments are a very good way of learning about the world, if you want to know what causes what, as well as knowing what tends to happen.

Remember John Snow from chapter 1, and his work on the transmission of cholera. In 1854, Snow stopped a cholera epidemic by having the handle of a public water pump removed. In a fictional modification of this case, suppose that things went as follows. Snow first observes that when the pump is not working, cholera rates go down. This might be because use of the pump causes the transmission of cholera (which suggests water contamination), but the situation might be different. Perhaps very cold weather tends to both knock out the pump and reduce cholera rates. This is another case with the general shape seen in figure 11.1. Now *A* is the working of the pump, *B* is cholera transmission, and *C* is cold weather. Perhaps not cold weather, but some other factor, knocks out both the pump and the epidemic together. Anything like this would be a *confounding* factor, or confounding variable, with respect to the apparent causal relationship between the pump and epidemics.

Snow can intervene to test this; he can first observe the situation and then remove the pump handle to see what ensues. This does not guarantee that he will learn the cause, but it makes a big difference. When he walks up and removes the pump handle, he cuts off a lot of the possible pathways by which some other factor might affect both the pump and the epidemic, because those other factors can no longer affect the state of the pump. Especially if he does this a number of times in different conditions, he can work out that the network is one where the arrow goes from pump to cholera. If you only do it once, like the actual John Snow, more uncertainty remains.

Here is an interesting real-life case from neuroscience (Marshall et al. 2006). When a person sleeps, it tends to help them consolidate memories from the day. Deep sleep also is associated with a particular kind of "wave" in the brain, a slow electrical oscillation at a rate of roughly one cycle per second. There has been debate about whether these measurable waves do something in the brain or are merely side effects of other processes. How can we find out? One way is to intervene and produce the waves, with electrodes placed on the skull, and see if memory improves. In an experiment, it did improve, and this supports the conclusion that these waves do have a causal role in consolidating memories. This does not yet tell us whether the waves have a relatively direct role in aiding memory, or the waves send the person into a form of sleep that includes a host of other changes that more directly affect memory (waves → other changes → better memory). Still, we have good evidence that the waves are not mere by-products, and instead play a causal role.

So observations are good in general, but experimental observations are often special, particularly if we want to learn causes. Given our earlier discussion of experiment in chapter 2, this seems a helpful insight. It also adds an element to the discussion of observation in chapter 7.

Interventionism is philosophically unusual in several ways, when compared to more traditional views of causation. The idea of an imagined intervention was used above. Do we need to purge the theory of any references to the imaginary or merely possible? In addition, interventionism is not an analysis of causation of the kind many people have sought, because it includes a kind of circularity. When we try to say what a causal relation is within this framework, we can't get away from using language that is causal itself. An intervention is itself a special kind of causal process, not merely an arrangement of events (Hitchcock 2019). People working on the interventionist research program have not been very worried about this. Sometimes their attitude is that they are developing a model that does not give the whole story but is definite progress anyway. Another option is to say that what we've learned is that the whole idea of causation is dependent on a certain point of view, the point of view of an *agent* who is reaching into a system (or imagining doing so) from the outside; this agent can't see themself as part of the system they plan to intervene on (Price 2005). In another sense, or from another point of view, any agent *has* to see themself as part of the world as a whole, part of a giant network, subject to causes as well as bringing about effects, at least unless that person has quite radical ideas about freedom. But perhaps there is a kind of mental "stepping outside," an unusual way of conceiving of things, that is part of causal thinking.

Further Reading and Notes

Wesley Salmon's *Four Decades of Scientific Explanation* (1989) is a survey of work on explanation between 1948 (the advent of the covering law theory) and the late 1980s. A good alternative discussion of causation and explanation can be found in Lewis (1986a). For unificationist theories of explanation, see Friedman (1974) and Kitcher (1989). Those ones are fairly advanced works, not easy to read.

Strevens (2004 and 2008) has developed a sophisticated account of scientific explanation that tries for a more unified view than my view here. In work on the biological sciences, an important development is the "new mechanist" view of explanation, which is notable for eschewing any important role for laws (Machamer, Darden, and Craver 2000).

Lewis discusses the Humean program in metaphysics in the preface to volume 2 of his *Philosophical Papers* (1986b). Beebee (2000) is a good discussion of the idea that laws "govern" things. Mitchell (2000) defends an interesting position on laws. For different approaches to causation, see Beebee, Hitchcock, and Menzies (2009). For the interventionist approach, see Pearl and Mackenzie (2018; an accessible introduction); Spirtes, Glymour, and Sheines (1993; an original source, much more technical); Pearl (2000; technical but clear); and Woodward (2003; focuses on the philosophical side). For the role of intervention in learning causes of disease, see Ross and Woodward (2016).

Chapter 12

Bayesianism and New Views of Evidence

12.1 *New Hope*

Through much of the twentieth century, unsolved problems about evidence hung over the philosophy of science. What is it for an observation to provide evidence for, or confirm, a scientific theory? Back in chapter 3, I described how this problem was tackled by logical empiricism. The logical empiricists wanted to start from simple, obvious ideas—like the idea that seeing many black ravens confirms the hypothesis that all ravens are black—and build from there to an "inductive logic" that would help us understand evidence in science. They failed, and we left the topic in a state of uncertainty and frustration.

Karl Popper was one person who could enjoy this situation, as he opposed the idea that the confirmation of theories is an essential part of science, and tried to maintain that science can be a rational process nonetheless. We can advance by knocking ideas out, even if we can never be confident that a theory is true. Other people came to accept rather vague holistic ideas about evidence, influenced by Quine. Somehow, observations can provide support for a whole "web of belief," even though it's not clear how a particular theory can be supported over others.

After discussing these ideas we moved on to a discussion of historically oriented theories of science, such as Kuhn's. These theories were not focused on the problem of confirmation in the way the empiricist views often were, but the problem did not go away. Even when this problem was not being discussed, it lurked in the background. If someone had come up with a really convincing theory of how evidence works, it would have been harder to argue for radical views of the kind discussed in chapters 7 and 8. The absence of such a theory put empiricist philosophers on the defensive.

The situation has now changed. Once again a large number of philosophers have real hope in a theory of confirmation and evidence. The new view is called *Bayesianism*. The core ideas of this approach developed slowly through the twentieth century; there was no breakthrough moment. But eventually these ideas started to look as though they might actually solve the problem. In this chapter I'll present the Bayesian view and some other new ideas about evidence as well.

Before we set out, I'll add that this chapter is probably the hardest in the book, and best approached by biting off small pieces at a time.

12.2 *Degrees of Belief*

At this point I will shift my terminology. The term "confirmation" was used by the logical empiricists and others, but recent discussions tend to focus on the concept of *evidence*. I used that term sometimes above, but from now on that will be the way things are set up. We want to understand evidence, for and against theories in science.

I'll start with a rather simple idea, and one that is some distance from the problem of evidence itself. Throughout this book I've talked often about *belief* in scientific theories and sometimes about *acceptance* of them. Most of the people I've been discussing think that in science, achieving certainty is practically impossible, so any belief in a theory (or acceptance of it) is always accompanied by awareness that you might be wrong. But the situation has generally been described as a yes-or-no decision. You accept a theory or you don't. You choose this theory or you choose another. You believe this hypothesis or not. However, doesn't this seem a bit artificial? A different way of setting things up is to talk about *degrees* of belief, or of confidence. You might have more confidence in theory X than you do in theory Y, where these are rival theories in the same field. You might have a high degree of confidence in one leading theory in the field, but also save a little for more speculative options. And as evidence comes in, you might get more confident about one theory and less confident about others.

The word "confidence" not quite right here. "Degree of belief" is better, but more technical. Sometimes people now talk about different amounts of "credence" you have for one theory or another. I am going to use "degree of confidence" and "degree of belief" as equivalents, choosing whichever phrase seems clearer in the context.

In philosophy, for a long time people treated a yes-or-no concept of belief as basic, and saw degree of belief as a complicated additional factor. There are ordinary beliefs, according to this view, and there are also partial beliefs, which are a special case. Sometimes it is hard to tell the difference between views that treat belief as a matter of degree and views that treat it as a yes-or-no matter but also accept that nothing is certain. A scientist might *believe that* theory *X* is the best supported theory, while also believing that other options are still on the table. If you set things up like that, this is not a degree-of-belief view, though it is getting close.

A few people saw things differently—they saw belief as always (or mostly) a matter of degree. The most important of these people was Frank Ramsey.

Ramsey was an astonishingly talented philosopher who died when he was only twenty-six. (He lived from 1903 to 1930.) In a career of only a few years, he did important work in several fields—philosophy, economics, and pure mathematics. His most significant work for us here is a paper called "Truth and Probability," written in 1926 but published after his death. That date takes us right back to the start of this book, the time of logical positivism. Ramsey was fairly well-known at the time in England and was in contact with some pivotal figures, including Wittgenstein. But the ideas he came up with that are most important for philosophy of science went unappreciated for a long time. Through the later part of the twentieth century, people began to see that Ramsey's ideas really might make a difference.

Here are the main ideas. We start out from the idea of degree of belief. It might seem that degrees of belief are very subtle features of a person—a strength of feeling, or something like that. No, said Ramsey; they are familiar and easy to get a handle on. A person's degrees of belief are revealed in their behavior. More exactly, a person's behavior is a consequence of their degrees of belief in various hypotheses along with their preferences or goals.

This is true of behavior of all kinds, but some kinds of behavior reveal degrees of belief with particular clarity. An example is betting behavior. We traditionally think of betting in contexts of games of chance, horse races, and so on. But all sorts of things we do can be seen as bets in a broad sense of the term. If there are clouds in the sky when you go out and you don't take an umbrella, you are betting that it won't rain. The Internet has also made it possible to place bets on a broad range of events, such as political elections. In "prediction markets," you make a prediction about a future event by buying "shares" in one or more of the outcomes. Suppose an election is coming up and you think that candidate *C* is pretty likely to win. You might pay 60 cents for a share in candidate *C* winning. If candidate *C* does win, you receive $1, and if the candidate loses you receive nothing. How much you will pay for a share of this kind is dependent on your confidence that candidate *C* will win. If you think it is almost certain that the candidate will, you will pay close to $1 for a share in that outcome, because even though the payoff is small you think you are extremely likely to get it.

Here is an example that Ramsey used. You are walking somewhere and are not sure if you are on the right road. How far would you walk out of your way to ask for directions? This depends on various things. It depends on how annoying it would be to later find yourself on the wrong road, and also on how much of a bother it is to go out of your way to ask for the directions. It might depend on other things, too. But one thing it depends on is how sure you are that you are *already* on the right road. The more sure you are that you are right, the less inclined you will be to go far out of your way to get directions. For Ramsey, all of life is a series of gambles, in which your behavior is a consequence of what you want to achieve and how confident you are that the world is one way rather than another.

So far we have the idea that everyone has degrees of belief in various ways the world might be, and these are revealed in our behavior. This seems a long way from science and evidence, though. How does this help? Before answering this, we'll shift gears again and move to what looks like a different topic.

12.3 *Understanding Evidence with Probability*

It often seems natural to express claims about evidence in terms of probability. We might say that seeing someone's car outside a party makes it very *likely*, or very *probable*, that he is at the party, for example. The use-fulness of talk of probability does not apply only to evidence. In the previous section, there were various places where you might have thought that it was natural to describe the situation in terms of probabilities—in terms of the chance that candidate *C* will win, or that rain will come.

Many philosophers have tried to understand evidence using probability. Here is an idea that lies behind many of these attempts: observational evidence can sometimes *raise or lower the probability* of a hypothesis. There is a formula from probability theory, a theorem that was proved several hundred years ago, that seems particularly helpful here. The formula is called "Bayes's theorem," after Thomas Bayes, the English clergyman who proved it.

Here is the formula in its simplest form:

$$(1) \quad P(h|e) = P(e|h)P(h)/P(e)$$

I'll also write it in a form that is more useful when thinking about evidence in science:

$$(2) \quad P(h|e) = \frac{P(e|h)P(h)}{P(e|h)P(h) + P(e|not\text{-}h)P(not\text{-}h)}$$

Here is how to read formulas of this kind: $P(X)$ is the probability of X. ("X" stands for some possibility, some way things might be.) $P(X|Y)$ is the probability of X conditional upon Y, or the probability of X *given* Y. Think of this as the probability that X has if we *assume Y*. The conditional probability does not depend on how probable we think Y is; we take Y for granted and ask how probable X is, given that assumption.

Looking back at formula (2) above, you should now think of h as a hypothesis and e as a piece of evidence. Then think of $P(h)$ as the probability of h measured without regard for evidence e. $P(h|e)$ is the probability of h *given e*, or the probability of the hypothesis *in the light of e*. Bayes's theorem tells us how to compute this latter number. As a consequence, we can measure what *difference* evidence e makes to the probability of h. So we can say that evidence *e confirms h* if $P(h|e) > P(h)$. That is, *e confirms* or *supports h* if it *makes h more probable* than it would otherwise be.

Bayes's theorem expresses $P(h|e)$ as a function of two different kinds of probability. Probabilities of hypotheses of the form P(*h*) are called *prior* probabilities. Looking at equation 2, we see P(*h*) and P(*not-h*); these are the prior probability of *h* and the prior probability of the negation of *h*. These two numbers must add up to one. The other probabilities are conditional probabilities, of the form P(*e*|*h*). These are often called *likelihoods*, or the likelihoods of evidence given various theoretical possibilities. In equation 2 we see two different likelihoods, P(*e*|*h*) and P(*e*|*not-h*). These need not add up to any particular value—a piece of evidence might be very probable (or improbable) if we assume *h*, and also if we assume not-*h*. (This special sense of the word "likelihood" can be confusing, as in many contexts "likelihood" just means "probability," without this being a conditional probability of a particular kind, as it is above.) Finally, P(*h*|*e*) is the "posterior probability" of *h*.

Suppose that all these probabilities make sense and can be known; let's see what Bayes's theorem can do. Imagine you are unsure about whether someone is at a party, as you arrive outside. The hypothesis that he is at the party is *h*. Then you see his car on the street out front. This is evidence *e*. Suppose that before seeing the car, you think the probability of his going to the party was 0.5. And the probability of his car's being outside if he *is* at the party is 0.8, because he usually drives to such events, though not always, while the probability of his car's being outside if he is *not* at the party is only 0.1. Then we can work out the probability that he is at the party *given* that his car is outside. Plugging the numbers into Bayes's theorem, we get P(*h*|*e*) = (.5)(.8)/[(.5)(.8) + (.5)(.1)], which is about 0.89. Seeing the car strongly confirms the hypothesis that the person is at the party.

This all seems to be working well. We can do a lot with Bayes's theorem, *if* it makes sense to talk about probabilities in these ways. It is common to think that it is not too hard to interpret probabilities of the form P(*e*|*h*), the likelihoods. Scientific theories are supposed to tell us *what we are likely to see*. The probabilities that are more controversial are the prior probabilities of hypotheses, such as P(*h*). What could this number possibly be measuring? And the posterior probability of *h* can only be computed if we have its prior probability.

Bayes's theorem was known, as I said, for a long time. And many philosophers writing about evidence, and also many statisticians, would have liked to use Bayes's theorem to make sense of how evidence works. But it was unclear whether this use could ever be legitimate, because to use Bayes's theorem we have to make sense of the prior probabilities of theories. Many people, especially statisticians, think that only the outcomes of repeatable events, such as repeatable experiments, can be described in terms of probability. The truth of a scientific theory about the world is not at all like this. If the prior probabilities of theories do not exist or are completely unknowable, then Bayes's theorem can't help us.

Perhaps what we need is a new interpretation of probability?

12.4 *The Subjectivist Interpretation of Probability*

Most attempts to analyze probability have taken probabilities to measure some real and "objective" feature of events. A probability value is seen as measuring the *chance* of an event happening, where this chance is somehow a feature of the event itself and its location in the world. That is how we usually speak about the probability of a horse winning a race, for example. There have been other interpretations, but many of them run into trouble. Ramsey, the philosopher I introduced earlier in this chapter, suggested an approach that initially looks rather strange, but turns out to be very impressive.

As I discussed above, Ramsey said that belief comes in degrees. Why not interpret these degrees of belief as probabilities? Mathematically that is possible. We can measure degrees of belief as numbers between 0 and 1, where 1 represents certainty that a claim is true and 0 represents certainty that it is false. Also, we can stipulate that if A and B are exclusive (can't both be true), then your degree of belief in A or B is your degree of belief in A plus your degree of belief in B. That is nearly all you need to interpret degrees of belief as probabilities (I will unpack this more thoroughly in a moment). So there can be a *subjective* interpretation of probability—an interpretation in which probabilities do not measure physical chances, frequencies of events, or anything like that, but measure degrees of confidence that some person has in a claim or hypothesis. If you are exactly balanced between thinking h is true and thinking it is false, then for you, P(h)=0.5. If someone says that the probability the horse Tom B will win its race tomorrow is 0.25, he is saying something about *his degree of confidence* that the horse will win. Another person might say that the probability is 0.8, and on a subjectivist interpretation of probability, she is just expressing *her* degree of confidence. They might both be rational.

So far Ramsey has done two things. He has argued that belief comes in degrees, and he has said that once these degrees of belief are interpreted numerically, we can treat degree of belief as a kind of probability (that is, the mathematics of probability theory can be interpreted as describing degrees of belief). Ramsey also accepted that there can be other kinds of probability; he just wanted to insist that one good way of thinking about probability is as degrees of belief.

Around the same time, another philosopher-mathematician, Bruno de Finetti, developed similar ideas to Ramsey, and in some ways he went further in arguing that the subjective interpretation of probability is better than others. (Below, I am going to use the word "subjective" for a kind of probability, and "subjectivism" for the philosophical view that we should generally (perhaps always) understand probability in a subjective way.)

The subjective interpretation of probability is not only important now in philosophy; it is central to decision theory, which has great importance in the social sciences, especially economics. Given the discussion earlier in this chapter, a particularly important thing subjectivism does is remove doubts about whether Bayes's theorem can be used to understand evidence. Above we looked at the problem of how to make sense of the prior probabilities of theories, such as P(h). There is no problem making sense of prior probabilities within a subjective interpretation of probability. P(h), for some person at a time, is just their overall degree of confidence at that time that h is true. When they see evidence e, their degree of confidence can change. The majority of philosophers who want to use Bayes's theorem to understand evidence today make use of a subjective view of probability—at least for applications of probability theory to this set of problems, and sometimes more generally. Some are subjectivists because they feel they have to be in order to use Bayes's theorem; others think that subjectivism is the only interpretation of probability that makes sense anyway.

In the rest of this section I'll look at some more details of the subjective interpretation of probability. These details are more technical, and it would be fine to skip to the beginning of the next section (12.5).

As I said, degrees of belief can be seen as probabilities—numbers between 0 and 1 that relate to each other in particular ways. Subjectivists also argue that if you have degrees of belief that are reflected in behavior, then you *have* to follow the rules of probability theory in how your degrees of belief in different things are related. The picture we have on the table now is one where each person at a time has a total network of degrees of belief—this is like a web of belief in Quine's sense, but with numbers attached to each belief that reflect the person's confidence. Subjectivist Bayesians argue that a person's network of degrees of belief is only "coherent" or rational when the person follows the standard rules of the mathematics of probability.

Those rules start from a set of axioms (most basic principles) developed by the Russian mathematician Andrey Kolmogorov. Here is a version of those axioms that is used by subjectivists:

Axiom 1: All probabilities are numbers between 0 and 1 (inclusive).
Axiom 2: If a proposition is a tautology (trivially or analytically true), then it has a probability of 1.
Axiom 3: If h and h^* are exclusive alternatives (they cannot both be true), then $P(h\text{-}or\text{-}h^*) = P(h) + P(h^*)$.
Axiom 4: $P(h|j) = P(h\&j)/P(j)$, provided that $P(j) > 0$.

(Bayes's theorem itself is a consequence of axiom 4. $P(h\&j)$ can be broken down both as $P(h|j)P(j)$ and as $P(j|h)P(h)$. So these are equal to each other, and Bayes's theorem follows trivially.)

Why should your degrees of belief follow these principles? Subjectivists argue for this with a form of argument traditionally called a "Dutch book." The idea is that if your degrees of belief do not conform to the principles of the probability calculus, there are possible gambling situations in which you are guaranteed to lose money, no matter how things turn out. These situations are very special; it is assumed that you are willing to bet on both sides of a proposition (or bet on all the horses in the race), at various different odds. In these situations, if the degrees of belief you have do not conform to the probability calculus, and you are willing to accept any bet that fits with your degrees of belief, then you will be willing to accept combinations of bets that guarantee you a loss.

Here is a simple example involving a coin toss. Suppose your degree of belief that the toss will come out heads is 0.75 and your degree of belief that the toss will come out tails is 0.75. Then you have violated the probability calculus because, by axiom 3, $P(heads\text{-}or\text{-}tails)$ = 1.5, which axiom 1 says is impossible. But suppose you persist with these degrees of belief and are willing to bet on them. More specifically, assume that if a bet is offered to you at odds that you think are fair, then you would be willing to accept *either* side of the bet. (I realize this is unrealistic, but let's see where it leads.) Suppose someone wants to bet with you on this coin toss. First, they want you to bet at 3:1 that the result will be heads. That means if the result is *not* heads, you pay them $3, and you get $1 if the result is heads. They also want you to bet at 3:1 that the result will be tails. You should accept both bets, because your fair odds for heads and for tails are both 3:1. (To go from a degree of belief p to odds of X:1, use $X = p/(1 - p)$.) But now you have accepted two bets that each pay worse than even money on the only two possible outcomes. So you are guaranteed to lose. If the coin lands heads, you win $1 on one bet but lose $3 on the other. The same applies if the coin lands tails. You have fallen victim to a Dutch book. If you want to ensure that no one could possibly make a Dutch book against you, you must ensure that your degrees of belief follow the rules of probability theory. This is a simple case, but more complex arguments of the same kind can be used to show that any violation of the mathematical rules of probability makes a person vulnerable to a Dutch book. Of course, one can avoid the threat by refusing to gamble at all. That is not the point. The point is supposed to be that the Dutch book argument shows that anyone who does not keep his degrees of belief in line with the probability calculus is irrational in an important sense.

12.5 *Bayesianism and Evidence*

Let us now go back to the problem of evidence. Earlier I discussed how, using Bayes's theorem, we might be able measure whether a piece of evidence raises or lowers the probability of a hypothesis h. Within some interpretations of probability, this might make no sense. But there is no problem within a subjective interpretation of probability. The "prior" probability of a theory, $P(h)$, is just the amount of confidence the person has in the theory before the next bit of evidence comes in. The probability $P(e|h)$ is the amount of confidence the person has in e, *on the assumption* that h is true. Both those assignments of probability (degree of belief) are made before e is observed. Then suppose e is actually observed. The person can calculate $P(h|e)$ using Bayes's theorem. They can say whether e supports h or not.

Things do not stop there; the story has an extra step. Now that e has been seen, the outcome of Bayes's theorem, $P(h|e)$, can be used as the *new* overall confidence the person has in h, which is applied when the next piece of evidence comes in. That is:

$$(3) \quad P_{new}(h) = P_{old}(h|e)$$

More informally, "today's posteriors are tomorrow's priors." A different (and more controversial) set of Dutch book arguments is sometimes used to argue that a rational agent should update her beliefs in accordance with formula (3). Bayesianism has to make special moves to deal with "old evidence," evidence known *before* its relation to a hypothesis is assessed. It also has a different formula to use when the evidence e is itself uncertain—this is important, because most philosophers accept that you can never be entirely certain about what you have seen.

We now have a model that covers not only the ways that evidence can support hypotheses, but also how a person's belief system can be repeatedly updated as new pieces of evidence come in. The whole package is quite impressive. But now let's start taking a closer look at what has been achieved, and what has not.

First, what kind of theory is this? It is a theory of rational belief changes by an individual. The overall picture is like this. You start out with some prior probabilities (degrees of belief) in all sorts of things, and you update as evidence comes in. Where you end up in this process depends on where you start. In describing evidence, we should not say things like "The probability that Darwinism is true is 0.9," or even "Given our evidence, the probability that Darwinism is true is 0.9." Instead, we should say something like "A rational person who starts out here . . . and sees this evidence . . . will end up with a degree of belief in Darwinism of 0.9."

This can be seen as a problem with the theory, a limitation on what it can handle. In standard presentations, a person is imagined as starting out with an initial set of prior probabilities for various hypotheses, where this initial set of prior probabilities is a sort of *free choice*; no initial set of prior probabilities is better than another so long as the axioms of probability are being respected. It is often seen as a problem that Bayesianism cannot criticize very strange initial assignments of probability. And where you end up after updating your probabilities depends on where you start.

That last point, though, is not so simple. Bayesians have argued that although the very first set of prior probabilities is freely chosen and might be weird, this starting point gets "washed out" by incoming evidence, so long as updating is done rationally. The starting point matters less and less as more evidence is taken into account.

This idea can be expressed as a kind of *convergence*. Consider two people with very different prior probabilities for hypothesis h but the same likelihoods for all possible pieces of evidence ($e_1, e_2, e_3, \ldots$), conditional on all the different hypotheses that are relevant. Now suppose also that these two people see all the same actual evidence, and this evidence, as it comes in, always pushes in the same direction—it favors the same hypotheses. Then, as long as neither person has a prior probability for *h* of exactly 0 or 1 at the start (and if they have zero probabilities for any other hypotheses competing with *h*, they agree on what these are), their degrees of belief in *h* will get closer and closer as the evidence comes in. It can be proved that for *any* amount of initial disagreement in their prior probabilities for *h*, there will be *some* amount of evidence that will get the two people, in these circumstances, to any specified degree of closeness in their final probabilities for that hypothesis. Even very large initial disagreements can be eventually overcome ("washed out") by the weight of evidence.

These "in the limit" proofs may not help that much. For any amount of evidence, and any measure of agreement, there is also *some* initial combination of prior probabilities such that this evidence will *not* get the two people to agree by the end. This fact and the "convergence" phenomenon are two sides of the same coin.

We might look at examples. Let's go back to my earlier example with the party and the car seen outside (section 12.3). A person had a prior probability of 0.5, and seeing the car outside raised that probability to about 0.89. Suppose that someone else has a prior probability of 0.2 for that person being at the party, but they agree with the first person about the potential relevance of seeing the car. Both people then see the car. If do the math again, we find that for the second person, P(*h*|*e*) is about 0.67. The two people are closer than they were before, though still some way apart. Now suppose they both see another bit of new evidence that they agree has the same likelihoods as the first—it has a likelihood of 0.8 given the person being at the party and 0.1 if he is not. After seeing the second piece of evidence, the first person's posterior probability goes to 0.98 and the second person's to 0.94. Now they are very close.

Some Bayesians have tried to work out a way of constraining the initial assignments of probability that Bayesianism allows, ruling out very strange ones. However, I think there is another problem with these arguments about convergence. They assume that when two people start with very different priors, they nonetheless agree about all their *likelihoods* (probabilities of the form P(*e*₁|*h*), etc.). But why should we expect this agreement? Why should two people who disagree on many things have the same likelihoods for all possible evidence? Why don't their disagreements affect their views on the relevance of possible observations? This agreement might be present, but there is no general reason why it should be. (This is another aspect of the problem of holism.) Presentations of Bayesianism sometimes use examples that involve familiar gambling games or sampling processes, in which it seems that there will surely be agreement about likelihoods even when people have different priors. But those cases are not typical of science. (Even in my party example above, you might be able to think of some factors that could lead people to disagree about the likelihoods.)

I'll move on to some other issues. Bayesianism is supposed to be a theory of *rational* belief change. Why should you follow the rules of probability theory when organizing your beliefs? Ramsey argued that if you don't follow these rules, you can get into trouble with your actions. Setting aside whether the argument given on this point (the Dutch book argument of section 12.4) is convincing, we can also ask another question. Is there any reason to think that managing our degrees of belief and our evidence in this way will help us get to the *truth*? That is a difficult question in many ways, and in the next chapter I will look at the idea of truth and how it operates in science. But as it happens, recent work has made some connections between the subjectivist Bayesian view of evidence and the goal of getting to the truth (Joyce 1998). Here is an example that conveys the main idea. Suppose you are a weather forecaster. Each morning you announce the chance of rain for that day (0.4, 0.9, or whatever). If it actually rains, we call that day a *one* day; if not, we call that day a *zero* day. You can then be given a score, as a forecaster, that depends on how big the gap is each day between your announced chance and the truth (1 or 0). If you said the chance was 0.4 and it did rain, your gap for that day might be 0.6–there is a range of slightly different ways of measuring the gap. A good forecaster is usually close to the truth, where this means that there is usually a small gap between their prediction and the way things turn out. The full story about how this relates to Bayesianism is complicated, but the bottom line is this: if you want the best score (for a wide range of reasonable ways of calculating that score), one thing you should do is follow the rules of probability theory.

Back in chapter 3 we looked at some classic problems involving evidence, Hempel's ravens problem and Goodman's "grue" problem. Bayesians have ways of dealing with each of these. I'll discuss Goodman's problem (from section 3.4) here. Suppose we are presented with two inductive arguments made from the same set of observations of green emeralds. One induction concludes that all emeralds are green, and the other concludes that all emeralds are *grue*, where an object is grue if and only if it was first observed before (say) 2050 and is green, or was not observed before then and is blue. The emeralds we have before us now are both green and grue. Why is one induction good and the other bad?

The usual Bayesian reply is that both inductive arguments are OK. Both hypotheses are confirmed by the observations of green emeralds. However, the "all emeralds are green" hypothesis will, for most people, have a much higher prior probability than the "all emeralds are grue" hypothesis. Then although both hypotheses are confirmed by the observations, the green-emerald hypothesis ends up with a much higher posterior probability than the grue-emerald hypothesis. That is the difference between the two inductions.

This does establish a difference, but why does the grue-emerald hypothesis have a low prior probability? Is anything stopping a person from having things the other way around—having a higher prior for the grue-emerald hypothesis? No. A person's prior probability for the grue-emerald hypothesis will usually be the result of much past experience with colors, minerals, and so on. But this need not be the case. Suppose you have never had a single thought about emeralds in your life, and you arbitrarily decide to set a higher prior for the grue-emerald hypothesis. Bayesianism offers no criticism of this decision, so long as your probabilities are internally coherent and you update properly. Is this a problem, or not?

Back in chapter 3 I also looked at the hypothetico-deductive (HD) view of evidence, and at some of its problems (sections 3.2 and 3.5). This is the idea that, roughly speaking, theories are confirmed when their observational consequences are seen to be true. I said back then that despite its problems, it's hard to believe that there is nothing right in the HD approach. The Bayesian framework makes sense of this situation.

Some, though not all, of the things to which a theory gives a high likelihood are its logical consequences (more exactly, they are consequences of the theory plus other assumptions). The cases that make the HD view look good can be accommodated by Bayesianism.

There is one more part of the Bayesian picture that I need to fill out; this is the relation between beliefs and actions. This might not seem obviously relevant to questions about science, but I think it is, and the reasons were discussed at the end of chapter 8 when we looked at "science and values" debates. The main ideas have been used implicitly a number of times already, but it is worth making them explicit. For a Bayesian, your degrees of belief are entirely determined by the evidence you see, along with prior probabilities. There is no place for giving some hypotheses the "benefit of the doubt," or some other preference that goes beyond what Bayes's theorem describes. When you work out what to *do*, you make use of your degrees of belief in all the ways the world might be, and consider (at least ideally) all the actions you might take and all the outcomes these actions might have in different possible states of the world. You choose the action that maximizes *expected value*. (This is often called "expected utility" or "expected payoff," but those terms suggest something inherently practical, and the model allows you to value whatever you like, whether or not it is practical.)

In a very simple case with just a few possible actions and two possible states of the world (represented by h_1 and h_2), you would go through the following calculation for each action:

(4) Expected value of action A_1 = $P(h_1)V(A_1|h_1) + P(h_2)V(A_1|h_2)$

And so on, for action A_2 and any others. You would then choose the act that gives you the highest expected value. Here "$V(A_1|h_1)$" looks a bit like a conditional probability, but it's not; it's a value or payoff—the value of performing A_1 in a situation where h_2 is true (e.g., carrying an umbrella all day when it doesn't rain).

In this framework, your choices about what to do take account of all the consequences your acts might have in different states of the world and how probable you think those states are. Your choices about what to *believe* are not affected by practical matters—only by the evidence, along with your priors.

People often say that the Bayesian model is not very realistic. In many ways I agree. Can we really expect people to have (even implicitly in their behavior) exact numbers for their degrees of belief, and to keep track of the likelihood of all possible bits of evidence, conditional on all the different hypotheses that might be true? No, this is clearly a long way from actual human thought processes. You might (and people often do) say that the Bayesian model is intended as a normative model, a model of how you *should* reason. In response, it might then be said that if some-one ever did follow the Bayesian rules exactly, they'd not get much else done in their life. Some of these features of Bayesianism have been made more realistic with further work—the framework can be developed with imprecise degrees of belief, for example—but a lot of it remains rather far from actual human thought. What a scientist might have, in many cases, is something like a solid confidence in one hypothesis, along with a sense of what the main alternatives are, and perhaps a lurking suspicion about one or more of these, where the confidence, suspicion, and so on are all matters of degree.

All in all, I think that Bayesianism is unrealistic in many ways, but it is a very good *model* of a couple of fundamental things in this area: degrees of belief, the updating of degrees of belief using evidence, and the relation between belief and action. Subjectivist Bayesianism is a good model of these things from both the descriptive and the normative points of view. It casts light on them by considering a streamlined, cleaned-up case.

I just described the Bayesian view (and did a few times earlier) as a "model." In using that word, I am making use of some associations that you might already have with the word "model" as opposed to "theory" in science. A model is often something that includes deliberate simplifica-tions. In the next chapter I'll look more closely at models, in this sense, and their role in science. Jumping ahead just a little, though, I think Bayesianism should be seen as a model of belief change—both of how belief change actually tends to work and how it should work when all goes well. The Bayesian model, even in more carefully formulated forms than the simple version discussed here, does not apply exactly to human minds, but it captures some basic relationships very well.

12.6 *Procedures and Experiments*

In this section I'll describe some other recent ideas about evidence. First, a theme that was neglected for quite a few years and has now been revived is the importance of inference via the elimination of alternatives—supporting one option by ruling out others. I will call this "eliminative inference." (It is sometimes called "eliminative induction," but that is another overly broad use of the term "induction.")

One reason this possibility was neglected for some time was that philosophers tended to assume that in science there is always an infinite number of possible alternatives to any given theory. If a theory has an infinite number of rivals, then ruling out any finite number of alternatives does not reduce the number of possibilities remaining. In principle, it might often be true that there is an infinite number of alternatives, but this might not be the end of the matter. Perhaps there are ways of constraining the *relevant* alternatives to a theory being considered, in which case we might be able to rule out most or all of the relevant alternatives. Importantly, scientists seem to give arguments of this kind all the time; there is no possibility that this is just a philosophical fiction. Given that, we should at least try to make more sense of it.

In very simple cases, eliminative inference can have a deductive form. Suppose you can decisively rule out all options except one. Then the inference can be presented as a deductive argument. (As always, such an argument is only as good as its premises.) There are two ways in which a nondeductive element can be introduced. First, there may be a less decisive ruling out of alternatives; maybe we can only hope to show that all alternatives except one are very unlikely. Second, we may be able to rule out most, but not all, of the alternatives to a hypothesis. Maybe as we rule out more and more of the alternatives to a given hypothesis, that hypothesis acquires a kind of partial support, even though some doubt remains.

The Bayesian framework can accommodate these sorts of cases, because Bayesianism always handles support in a *comparative* manner. For one hypothesis to gain some credibility, at least one other hypothesis has to lose some. Even in the simple cases I went through above, there was competition between *h* and *not-h*. Suppose there are three possibilities. Then formula (2) will look like this:

$$(4) \quad P(h_1|e) = \frac{P(e|h_1)P(h_1)}{P(e|h_1)P(h_1) + P(e|h_2)P(h_2) + P(e|h_3)P(h_3)}$$

At this point, we reach one of the main problems with attempts to learn about the world by eliminative inference, a problem that can be represented clearly in the Bayesian framework. How can we know that we have considered all the relevant alternatives? It can be argued that scientists constantly tend toward overconfidence on this point (Stanford 2006; see also section 10.6). Scientists often think they have ruled out (or rendered very unlikely) all the feasible alternatives to their preferred theory, but in hindsight we can often see that they did not do so, because we now *believe* a theory they did not even *consider*. Progress often requires putting totally new options on the table.

In formula (4) above, this problem shows up in the assumption that the three theories listed are the only options. If they do not really exhaust the possibilities, and there is an h_4 lurking in the background, the Bayesian inference will go astray. (Actually, it's not quite as simple as that, because the Bayesian view is, again, a theory of management of degrees of belief. If you *think* that h_1–h_3 are the only alternatives, you can use Bayes's theorem as usual. But if someone then comes up with a completely new option, you will have to do it all again.)

In a Bayesian model, this situation is often represented by supposing that a person has a couple of definite alternatives in mind (h_1, h_2, h_3) along with a "catch-all" or "cover-all" hypothesis that just says: *something other than h_1–h_3 holds*. And how can you work out how likely evidence *e* is if *something* other than h_1–h_3 holds? Can you give a number to $P(e|$something else I haven't thought of)? If you can't, you can't use Bayes's theorem to update your beliefs.

The next theme I'll look at is another general idea about what makes an observation into evidence. Specifically, the idea I will look at is that we should analyze evidence by focusing on *procedures*. The ideas that arise here can in many cases be combined with Bayesianism, though they might also provide parts of an alternative picture.

The guiding idea is that if an observation provides support for a theory, in many cases (not all), that support exists because of the procedure that the observation was embedded within. Not all procedures must be explicit, planned tests or experiments; some can be more informal. An important source for this procedural approach (as I'll refer to it) is Hans Reichenbach, who was associated with the logical positivists but had somewhat different views about evidence. My version of this idea uses the idea of reliability; a good procedure is one that has the capacity to reliably answer the questions we put to it. The overall orientation here is also naturalistic (chapter 9), in a broad sense.

I'll illustrate this view by looking at a particular kind of procedure employed in science: using random samples to make inferences about the characteristics of a larger population. This is the kind of procedure involved when we use a survey to find out (for example) how many teenagers smoke cigarettes. In some ways, this is the closest thing to a scientific home for the traditional philosophical picture of inductive inference. If we approach some of the standard philosophical problems around induction from a procedure-based point of view, it makes a difference. Let us look again at the two famous puzzle cases discussed in chapter 3, the ravens problem (section 3.3) and Goodman's "grue" problem (section 3.4).

Here is a recap of the ravens problem. If generalizations are confirmed by their instances, and if any observation that confirms *H* also confirms anything logically equivalent to *H*, then it seems that a white shoe confirms the hypothesis that all ravens are black. After all, the white shoe is a nonblack nonraven. So it is an instance of the generalization "all nonblack things are nonravens," which is logically equivalent to "all ravens are black."

In most philosophical discussions of this problem, there is little attention paid to how the observations of ravens (and shoes) are being collected. Of course, the whole example is unrealistic. A biologist would not try to learn about bird color simply by generalizing from observed cases. But imagine that a biologist is doing something like this. Rather than relying on casual observation, though, the biologist uses a statistical method. Let us distinguish two questions about ravens:

The general raven question: What is the proportion of blackness among ravens?
The specific raven question: Is it the case that 100 percent of ravens are black?

Questions of this kind can be reliably answered using samples from a larger population, if we have a sample that is *random* and of a reasonable *size*. Statistical theory will tell us exactly how large a sample we need in order to get an answer with a desired degree of reliability (here "size" of sample means absolute size, not size in relation to the size of the overall population). How might we collect an appropriate sample?

The most obvious approach is to collect a random sample of ravens and record the birds' colors. A sample of this kind can be used to answer both the specific and the general raven questions, using ordinary statistical methods. So far, so good.

But now consider a more unusual approach. Suppose we could collect a random sample of nonblack things and record whether or not they are ravens. This method will be useless for answering the general raven question. Interestingly, though, it *can* be used to reliably answer the *specific* raven question. If there are nonblack ravens, we can learn this, in principle, by randomly sampling the nonblack things.

Now that we are imagining unusual sampling methods, there are two others to consider: collecting a sample of black things, and collecting a sample of nonravens. Neither of these can be used, without further assumptions, to answer either of the raven questions. Knowing what proportion of the black things are ravens does not tell us what proportion of ravens are black, and a sample of nonravens is of no use either.

So far we have distinguished between some procedures that can, and some that cannot, answer our questions about ravens. Now we can look at the role of particular observations. Consider a particular observation of a white shoe. Does it tell us anything about raven color? It depends on what procedure the observation was part of. If the white shoe was encountered as part of a random sample of nonblack things, then it is a piece of evidence. It is just one data point, but it is a nonblack thing that turned out not to be a raven. It is part of a sample that we can use to answer the specific question (though not the general question), and work out whether there are nonblack ravens. But if the very same white shoe is encountered in a sample of nonravens, it tells us nothing. The observation is now part of a procedure that cannot answer either question.

The same is true with observations of black ravens. If we see a black raven in a random sample of ravens, it is informative. It is just one data point, but it is part of a sample that can answer our questions. But the same black raven tells us nothing about our two raven questions if it is encountered in a sample of black things; there is no way to use such a sample to answer either question.

In situations of this kind, an observation is only evidence for a hypothesis—even if that hypothesis is a simple generalization—if it is embedded in the right kind of procedure. This does not apply in the case of deductive relationships. A black raven refutes the hypothesis that no ravens are black, regardless of the procedure behind the observation. But deduction is special.

This approach to the ravens problem is a more elaborate version of the idea, discussed in chapter 3, that order of observation is important. But what is important is not order, but procedures.

I turn now to Goodman's problem (which I reintroduced in the previous section of this chapter). This one is harder, because I think the "grue problem" combines several different problems together (including the problem of simplicity). But I will present part of an answer.

Let us continue thinking about inferences made from samples using statistical methods. These methods can be powerful, but they can only be used when some assumptions hold about the testing situation. One situation where these methods cannot be used, in their simple forms, is when the act of observing or collecting the data changes the particular objects being observed, in a way that is relevant to the question being asked. In some cases we might overcome the problem by taking into account the effects of our data collection and compensating for this fact. But special measures of some kind will be needed.

Now consider Goodman and his emeralds. Again, the philosophical literature has chosen a bad example here, but suppose we are making inferences about all emeralds by observing a random sample. This method would encounter a problem, of course, if the act of collecting or observing individual emeralds changed their color. In such a case, a simple extrapolation from the color of the sample to the color of the unobserved emeralds would be unreliable. That problem is obvious. But there is a less obvious connection between this case and the grue problem.

First, we should remember that a grue object is not one that changes color at some special date. A grue object is one that was either first observed before 2050 and is green, or was not observed before 2050 and is blue. With that point clear, think about a sample of grue emeralds that we might have collected.

To keep things simple, suppose that *all* of our previously observed emeralds are in the sample. So we have a big pile of emeralds, all of them grue. The act of putting them in the sample did not physically change them, but something related is going on. If those particular emeralds had not been observed before 2050, they would not have been grue. After all, those emeralds are green, and anything green that was never observed before 2050 does not count as grue. So the grueness of an object depends, in an odd conceptual way, on whether or not the object has been observed before a certain date. Putting it loosely, the emeralds in the sample were *affected*, with respect to their grueness, by the fact that they had been observed before now. That means we cannot extrapolate grueness from the sampled emeralds to the unsampled ones. We cannot do the extrapolation because the observation process has interfered, in an unusual way, with the characteristics of the objects in the sample. This problem does not appear if we want to extrapolate greenness from a sample of emeralds; it only appears if we want to extrapolate grueness.

The grue problem (or this aspect of the grue problem) is a philosophical relative of a familiar problem in statistical methodology in science. It is akin to what would be called a *confounding variable* problem. In a way, Goodman's term "grue" turns observation (or sampling) itself into a confounding variable. This shows, again, the importance of procedures when thinking about evidence.

The methods described here are not the only ways to investigate generalizations. There are least two "paths" from observed to unobserved cases in a collection of objects. One is random sampling, but many collections can't be randomly sampled. One can often learn about unobserved cases by making inferences from the internal structure or causal properties of the observed ones (as we do with actual emeralds). Then randomness plays no role, and more weight falls on the nature of the collection or kind we are studying (see section 10.7).

I'll finish the chapter with one more set of ideas. A theme that has come up several times in this book is the role of experiment, as opposed to other kinds of observation. Are experimental observations superior to others? If so, what is the dividing line between the two? Is looking up at the sky every night at midnight an experiment, or do you have to manipulate the system you are studying?

I think there is a lot still to work out here, but some ideas, and perhaps some progress on the topic, follow from what I said just above, especially in combination with some ideas at the end of the previous chapter. First, the very idea of a procedure that is followed in collecting data takes us some way from a purely passive or happenstance collection of observations. Above, in the discussion of the ravens problem, we saw that the same observation (of a black raven) could have a very different role as evidence according to the procedure the observation was embedded in. Collecting a random sample of black ravens is a bit like an experiment, but a very minimal one. We need not interfere with the ravens at all. It's a bit like pointing a telescope each night at the sky at the same time. Still, as we saw above, it can make a big difference that we are collecting a raven sample rather than a sample of black things, or not collecting a random sample of anything at all. So there is one difference we see here between observations collected haphazardly and observations collected according to a procedure. We can answer questions with the second that we can't answer with the first. There is one kind of superiority for "experiments" that reflects the special role of procedures, and no more than that. This superiority is seen in experiments that do not involve intervention in the objects being studied.

However, there is also the quite separate phenomenon we saw at the end of chapter 11, which has to do with intervention and learning about causal relationships. In many cases, the best way (sometimes the only way) to learn whether *A* causes *B* or some other cause *C* is responsible for the association between *A* and *B* is to do an experiment that includes an intervention—take the handle off the water pump, stimulate a person's brain through the skull, give them a new drug and see what happens. Intervention enables us to control for a lot of potential "confounding" factors. Here we see a role for experiment that requires more than just having a procedure; we have to *do* things to the system being studied.

So there are two separate points to be made, given the last two chapters, about the evidential importance of experiment. One applies to experiments that involve procedures for gathering evidence, whether or not there is intervention. Another involves experiments in a richer sense, where we learn about the world by interfering with its workings. There is probably more to the situation than this, too; by means of experiments in the richer sense, we can isolate objects, and we can also reveal hidden behaviors of objects by putting them in new settings. The more general conclusion I want to draw, though, is that there are some advantages to doing experiments that merely or mostly involve the existence of a procedure, and other advantages that involve more.

Further Reading and Notes

On Bayesianism, I find Howson and Urbach's *Scientific Reasoning: The Bayesian Approach* (now in its third edition, 2005) very helpful. John Earman's *Bayes or Bust?* (1992) contains some important ideas, but it is hard work. Hájek and Joyce (2008) cover these ideas in a way that preserves continuities with Hempel's project. Bayes's own "Essay towards Solving a Problem in the Doctrine of Chances" was published posthumously in 1763. For an example of a prediction market, see website [5].

Brian Skyrms's *Choice and Chance* (2000) is a classic introduction to probability and induction. See de Finetti ([1931] 1989) for his version of subjectivism. Misak (2020) is a biography of the remarkable Frank Ramsey. Resnik (1987) is a particularly helpful introduction to decision theory, subjective probability, and Dutch books. Richard Jeffrey's *The Logic of Decision* (1965) is a classic treatment of several of these topics and includes a formula for updating beliefs when the evidence is itself uncertain. Salmon (1990) is a good paper about unconceived alternative hypotheses and how Bayesianism relates to Kuhn's ideas. For a Bayesian treatment of the ravens problem, see Rinard (2014). Forber (2011) reconceives eliminative inference.

Reichenbach's main discussion of his procedure-based approach is in *Experience and Prediction* (1938). *The Rise of Scientific Philosophy* (1951) has a simpler version. In the first edition of this book I called my approach *procedural naturalism*. I think the link to naturalism is optional, though. (That phrase also has a more established meaning in another part of philosophy, jurisprudence.) Mayo's approach to evidence (e.g., 1996) is also a procedure-based one, focused especially on statistics. Jackson (1975) proposed a solution to Goodman's problem of roughly the kind I outline here, without tying the solution to statistical methods. I follow up these ideas in more detail in Godfrey-Smith (2003 and 2011). Norton (2003) argues for a related "material" theory of induction, and Okasha (2011) discusses the ravens, Bayesianism, and experimental intervention in a single philosophical cocktail.

Chapter 13

Truth, Simplicity, and Other Problems

13.1 *The Problem of Truth*

This chapter will cover four topics, one in detail and three in briefer sketches. The first topic is *truth*. What is it? What makes a theory true? What makes truth a good thing, if it is a good thing? Continuing with similar themes, I'll then look at a particular kind of scientific work that can have an unusual relationship to the goal of truth. This is model-building, an approach to theorizing that features deliberate simplification and other departures from reality, apparently in the service of understanding. After that, I will look at *consensus* in science, at what it means for a field to decide that a question is settled. Lastly we'll turn to a famous principle that is often invoked in science and philosophy to defend one theory over another. "Occam's razor" says we should prefer simple theories to complex ones, whenever we can. If so, why?

Truth has been playing a tacit role all the way through the book, all through the discussions of evidence, belief, and justification. It seems that our goal is truth—finding and accepting true theories—and evidence is what we use to get to that goal, if we can. Truth itself has been mostly in the background in these discussions. Now we'll look at it explicitly.

It is remarkable how hard this problem is. Every step seems uncertain, and seems to risk taking us down a wrong (false!) path. Truth really is a thicket. I won't try to take things too far in this discussion, but I will introduce the problem and go some distance forward.

Consider the perspective of a complete outsider looking at science and its place in human life. We might imagine a Martian coming down to observe us. The Martian sees that we talk a lot and write things down. We make inscriptions and type on keyboards and screens. People fiddle with complicated machines that process exotic chemicals and type things afterward. Then others often go off and act in a way that appears to be guided by what has been written down. Those actions sometimes have remarkable results; people build bridges that stay up and make drugs that cure infections. It seems that some of the inscriptions, diagrams, and computer files have something about them that enables good bridges and effective drugs to be made, while other inscriptions (etc.) lead people astray. The Martian might ask: what is the feature that enables this?

In one way this thought experiment is misleading. If Martians can get to Earth to watch us, they must have been doing something similar themselves—must have their own scientific theories in order to build the technology that got them here. They must communicate and build on the work done before them using records of their work. They must feel at home, in some ways, with the behaviors they are studying in us. Still, it's worthwhile to think about the question from this point of view.

I said there seems to be a sort of power in some of the sentences, charts, and equations to lead to drugs and satellites that work. But it might be an error to put the power so much in those sentences, equations, and so on themselves. Clearly a great deal depends on the *interpretation* of those things. If a person is given the sentences alone, without knowing how to interpret them, nothing useful can be done. Technological successes seem to require a pair of things: the right representations and an ability to interpret them.

Somewhere in here is the distinction between true theories and false theories, or between more accurate and less accurate theories. Or, perhaps, between theories that are closer to the truth and those further from it—a distinction that still gives truth a role.

Another way to put the general question here is like this. What is the right way to present what we take to be the *achievement* reached, when theoretical science goes well? We might try to understand the goals and achievements of science while working as little as possible with the idea of truth. We might say the goal of theories is to predict events only, not represent the workings of nature. That does not entirely remove the role of truth. To make a prediction is to say something about what we will see or hear, and to make a good prediction is to say something true about what we will see or hear. The idea of truth has not disappeared yet, though it might eventually. But for now, let's try to make sense of truth rather than getting rid of it.

13.2 *Correspondence, Coherence, and Usefulness*

I'll make my way in by looking at some very general ideas about truth—not ideas about science, but about language in general and related topics. In particular, I'll start with some standard ideas that are all, in a sense, direct attempts to say what truth is—what it is in everyday cases as well as scientific ones. These views try to describe a special feature of some beliefs or representations that makes them true. The next section will then take a different tack, one that I think is better.

There is a menu of standard philosophical options for making sense of truth. These include correspondence views, coherence views, and pragmatist views.

Correspondence views pursue the idea that truth is a match of some sort between a belief or theory and the way things are. Perhaps truth is a sort of picturing—a lot of people have thought there is something in this idea, not just for things that obviously look like pictures (maps, diagrams), but generally. Scientific theories might be, in some abstract sense, pictures or maps of the world.

This approach is tied up with the problematic notion of a *fact* (a concept that caused some trouble in section 7.4). Correspondence is often said to exist between beliefs or theories, on one side, and the facts, on the other. People sometimes talk of a fact as a piece of the world itself, but sometimes a fact is seen as something like a true sentence. The word unhelpfully occupies both sides of the relationship we want to understand—the relationship between representations and the world they are aimed at. I will mostly avoid the term "fact" and look instead at the idea of truth as correspondence with how things are, with the world itself.

Correspondence views are often seen as a natural place to start, but hard to defend under pressure. Sometimes they are attacked with an argument like this: we can never hold up our beliefs and the world itself, as separate things, and compare them to see whether there is a correspondence between them. We can't step outside our beliefs, categorizations, and perceptions. So even if a correspondence relation existed between some beliefs and the world, it could have no role in our lives.

I don't think this is much of an argument. There might be a relation between a belief and the world that is a sort of match or picturing, even if you could not, in any direct way, check to see if it is there. We get some evidence about the existence of this relation from our predictions, but can never be sure. Beliefs that correspond to the way the world is might also tend to produce more successful actions, even though you might often not know whether an apparent success really counts as one, because you don't know how things would have gone if you had behaved differently.

A different problem is that it is hard to see what this correspondence relation might actually be. "Picturing" is a mere metaphor, one applicable in simple cases but not helping much with most cases, especially in science (Rorty 1982). This, I think, is a more serious issue.

A second standard option is that truth is a kind of *coherence* between beliefs themselves—a form of harmony or stability within a belief system. This initially seems rather far from truth, because coherence is a relation between one belief and others, not between beliefs and a world they are aimed at. But there are ways to motivate coherence views. Look more closely at the idea that truth is a goal, something we search for and value. How do we do that? Coherence is what we can actually work on in our belief systems, something we can aim for. It is also something whose presence we can discern for ourselves (though not infallibly). In these respects, coherence seems to fit the role of truth in our everyday lives.

Isn't observational data something that gives a belief system contact with the world beyond it, hence a constraint that goes beyond "coherence"? In reply, it can be argued that an observation plays this role only when it has been described and recorded, and then it is just another part of a web of beliefs that may either be coherent or not.

Coherence is a matter of degree, while truth, it seems, is not. Belief systems can be either more coherent or less so, but beliefs are just true or false. In response, we might modify the idea of truth. Maybe it is a matter of degree, too.

A third family of views about truth comes from the American pragmatist movement of the late nineteenth and early twentieth centuries. During this period, pragmatism developed two main ideas about truth. C. S. Peirce suggested in 1878 that what we mean by "the truth" is what would be believed at a sort of ideal limit of scientific investigation—what would be believed in a consensus that would be reached, or at least approached, if inquiry went on indefinitely.

A second theme in pragmatism uses not the link between truth and evidence, but the link between truth and action. True beliefs, it seems, are ones that tend to lead us to successful actions. And maybe that is the heart of the matter; maybe *all we* mean by "true belief" is a belief that will work well for us, one that helps us achieve our goals when we act on it.

This last option was criticized heavily after being proposed by William James in the early twentieth century (1907). Though there does seem to be a link of some kind between truth and successful action, it's hard to believe the relationship is as simple as this. Something can be a little bit useful, or useful some of the time. Sometimes James accepted this and said that a belief can be a bit true, or true for me but not you, or true in some contexts. But the whole concept of truth seems to include the idea that a belief could work well for someone despite being false. It seems much more plausible to have a view in which truth is some sort of relation between belief and the world that produces or aids successful action. A theory is not true because the bridges we build with it stay up—that does not make the theory true. Instead, the truth of a theory (in materials science, for example) is something that is partly *responsible* for the fact that bridges we build using that theory tend to stay up, or are more likely to stay up than others. The pragmatist view seems to have things the wrong way round.

13.3 *An Indirect Approach, via Ramsey*

A better road is to approach truth in a slightly more indirect way. I am influenced here by Frank Ramsey, the same person who gave us degrees of belief and the subjective interpretation of probability in chapter 12.

Here is the main idea. Suppose we know exactly what is involved in someone making a claim, an assertion, saying that the world is a certain way. When I say "exactly what is involved," I mean that our Martian could come down and say, this person here is making a claim that such-and-such—that the cat is asleep or that photons have no mass, or whatever. Once we understand this, making sense of truth is easy. The claim made is that the cat is asleep. Well, is the cat asleep? If it is, then the claim is true. If not, the claim is false.

Aristotle, over two millennia ago, said about truth: "To say of what is that it is not, or of what is not that it is, is false, while to say of what is that it is, and of what is not that it is not, is true" (*Metaphysics* 1011b25). That is often seen as a statement of a correspondence view, and in a way it is—in a minimal, low-key sense of "correspondence." We have a claim that X is how things are in the world, and in the world X is indeed how things are; what is claimed corresponds to the way things are. But correspondence here does not involve picturing, or anything like that. Instead, it is just a simple relation between what is said and the way things are.

We can think about making a claim, or offering a hypothesis, with the idea of a *satisfaction condition*. (My use of this phrase here modifies some more standard and more technical uses.) When you make a claim, what you say has a satisfaction condition, a condition that might or might not hold. *Photons having no mass* is an example—it's a way things might be. Truth is what you have when the satisfaction conditions of what you say are actually satisfied. We then don't need "picturing" for truth (unless picturing is no more than this). The hard part of the problem becomes explaining what it is to make a definite claim about how things are, what it is to say something that has a satisfaction condition. Ramsey's view is not that making sense of truth is easy; it is easy once we have made sense of assertion, and that first task is not easy.

The philosophical debate about truth includes another view called *deflationism* or *minimalism* (Horwich 1990). This is the view that there is no philosophical problem of truth at all. To say "X is true" is just to say X. Truth is not a special and hard-to-analyze relation between words and the world. The word "true" is just a tool we use to say some things in a convenient form. (Saying "everything Bob just said is true" is like repeating everything Bob said, and also giving him credit for saying it.)

The term "deflationism" indicates that we deflate the problem; rather than finally finding the hidden nature of truth, we show that this was never really a problem.

Ramsey's view is sometimes seen as early statement of deflationism. In a way it is, as he does deflate the problem of truth itself. But this is only because he thinks the hard part is explaining what it is to make a claim or assertion.

Again, the goal here is a theory that would work for someone looking at us entirely from outside, like the Martian observer. To appreciate the problem, we need to not take our own meaningful speech for granted. If we do take that speech for granted, there indeed seems to be no problem at all here: "Of course I know what it is to assert things, and so do you. When we speak, we usually mean what we say. If I say 'photons have no mass' and they indeed don't, then I said something true. No problem." The Martian can't do that. The Martian comes down and sees sounds, inscriptions, and files that have no obvious meaning, from a Martian point of view, but some of which seem to have unusual power. The Martian wants to understand how all this works. So the Martian (as well as we ourselves, trying to look on from outside) needs to start afresh: are some of the marks made by scientists sentences with satisfaction conditions? For example, do some of them claim that photons have no mass? Some philosophical possibilities we've looked at earlier reappear here. Perhaps it is a mistake about the nature of language to suppose that people say things that have satisfaction conditions tying those claims to definite situations in the world. Perhaps it is only possible to talk about our experiences, or perhaps we can only "aim" our assertions at vague clumps of the world rather than at photons or cats in particular.

What we need next, then, is a theory of whether and how it is possible to make claims about specific aspects of the world, of the kind relevant to science. We want a theory that covers not just ordinary language but mathematical formulas, diagrams, and so on—I will call all these "representational devices." There are some detailed philosophical theories in this area, mostly covering language rather than other kinds of representation, and these are sometimes extended to science (Kitcher 1993; Devitt 1997). I don't think any of these theories work very well. But we do know roughly what is involved. We have, first, a mass of practices of creating representations and using them—saying things and listening, typing symbols and reading them. These practices of making and using representations are embedded in various other behaviors. Second, there are causal relations that link these practices and the representations themselves to various other parts of the natural world, sometimes with the aid of elaborate instruments. And third, there is commentary that people engage in, using one representational device to say something about another: "When I use the term 'species' I mean. . . ."; "Let *m* represent momentum. . . ." The theories of meaning and representation that people have come up with tend to emphasize one or the other of these— the making and use of signs, or causal relations between signs and other objects, or commentary on them—as *the* factor that determines what they mean. Maybe, instead, a good theory must give a role to all of these.

I won't try to give that theory here, but will assume that the upshot of all these facts is that by using our representational devices, one thing we can do is express claims about how things are in the world at large. And once we have that, the existence of truth and falsity follows. So does correspondence, in a low-key sense. If you say that *X* is how things are, and things actually are that way, that is a kind of correspondence. Coherence, which was also discussed in the previous section, has a less intimate connection with truth. It is, at best, an indicator of truth. The same applies to usefulness.

What becomes of the debates about whether truth is relative or absolute, whether we each have our own truth, and so on? Some of these debates must be seen as confused, for one reason or another—sometimes because of conflation of what is true with what merely seems true, sometimes because of other things. But there are genuine questions here, too. The idea that truth is relative might be understood as involving a kind of relativity in the world itself—the world being this way from your perspective, another way from mine. If there is no "objective" reality—no way the world is, regardless of perspective—then truth also will be relative to perspective. Truth will be correspondence to the perspective-dependent objects. This is a view that might be defended, though I think it is a mistake. (Another possibility is that the relativity of truth, or something like it, might stem from the other side of the relationship, the side of assertions or representations. If it's not possible to say something with definite satisfaction conditions, and "what is said" is merely a matter of interpretation, then whether what was said is true or false will also be a matter of interpretation.)

If all this gives us some handle on the nature of truth, why is truth a good thing? Why is it a goal for science and other investigation? I'll discuss two ideas here, one that is general and one that connects more to science.

First, let's look at some general features of the connection between belief and action. In a host of everyday cases, we work out how we think the world is, and then use that understanding of the world to help us work out how to act. We act in ways that make sense—that should work—given how we *think* the world is. If the way we *think* the world is turns out to be the way it *actually* is, then our actions will turn out to be well chosen. If we are wrong about how the world is, we will act in ways that are poorly adapted to that situation. Or, at least, we will act in ways we *think* are poorly adapted to that situation—we might be wrong about which actions are good ones in a particular state of the world, as well as wrong about how the world actually is. The "expected value" model of action that I discussed in the previous chapter is a more detailed formulation of this idea, one that takes into account the fact that we are typically uncertain about how the world is, but have more confidence in some possibilities than others.

All this tells us something about why truth is a goal. We continually act on the assumption that the world is one way rather than another way. True belief is, among other things, what you have when your behaviors are based on assumptions about the world that do hold.

This makes some sense of the practical value of truth. Many would say that truth has a value that goes beyond any practical usefulness—beyond any links to action. I am sympathetic with that view, though I am not sure whether it goes beyond a personal choice.

The second idea I will look at is more specifically concerned with science. Science does not only make claims about how things are. We also develop elaborate ways of expressing scientific claims—ways of categorizing things, ways of organizing ideas. Science introduces special means for representation, including mathematical formalisms, schemata for diagrams, and the structures within computer programs. These are not themselves assertions or claims; they are ways of making assertions and claims, tools for representing things. They do sometimes embody or assume claims about how things are. The periodic table of chemical elements, for example, is an organizing device that embodies claims about the chemical world.

In some of these cases, an organizing format is devised that has a map-like relation to the part of the world it is aimed at dealing with. The periodic table is like a map of chemical reality, with locations on the table corresponding to locations in a space of chemical similarity. It might be tempting to push this idea further and say that many, or all, special-purpose symbolisms introduced in science are maps, in an abstract sense, of some part of the world. (More precisely, they are *supposed* to work as maps; like any map, they may fail). Then the usefulness of these devices might be understood as similar to the usefulness of more familiar kinds of map, and this might be a further role for the idea of correspondence in explaining how scientific representations work.

I think there are some cases like this, but the last move I made just above was probably an overstatement. Often, the power of an organizing format or representational system does not come from its operating as a map, even of a subtle kind, but from something else. Representational systems in science often enable the transformation and manipulation of ideas, and that is a power they have that does not require that they "map" anything. An example is provided by mathematical symbols. These are means for making assertions, but they also facilitate the manipulation of information, by substitutions and transformations. In this way they enable the continual exploration and extension of ideas.

13.4 *Models*

The simplest way to think about a scientific theory is as an organized collection of claims about the world, claims that might be true. In some parts of science, things seem to be set up differently. I'll use an example to introduce this idea, the hawk-dove model from behavioral biology, first developed by John Maynard Smith and George Price in 1973 and modernized a little here.

Assume that a population contains two types of individuals, *hawks* and *doves*. They meet randomly in pairs, and engage in contests over resources. When a hawk meets a dove, the hawk wins the resource, with no injuries on either side. When a dove meets a dove, it's random who wins, and again there are no injuries. When a hawk meets a hawk, it's random who wins, and the loser is injured. Assume a process of natural selection operating in the population, so that differences in reproductive success are a consequence of success in these interactions, and assume that each behavior is passed on in reproduction, where reproduction is asexual (everyone just has one parent). Then as long as the cost of injury is greater than the value of the resource at stake, the population will tend to reach a stable mixture of hawks and doves. Each type does comparatively well when it is rare, and badly when it is common.

That is interesting, but no population actually works like this—with two types of individuals who all behave the same way within their type, encounter each other always in pairs, and so on. Some populations might *approximate* situations like this, though, and we seem to have gained an insight here into the consequences of having more and less aggressive individuals around in a biological setting.

In situations like this, what we seem to do is first specify a structure, an arrangement, and see how it will behave. Then we use those results to draw conclusions about the actual world. An "arrangement" of this kind (as in the hawk-dove model) looks like a fiction, an imaginary system that can be compared to reality. It is similar to the real world in various ways, but easier to comprehend. This combination—close to reality but simpler—makes it informative. We can learn things about a complicated real system by learning about a simpler, imaginary version.

Figure 13.1 shows how this works (the figure is modified from one used by Ron Giere in his book *Explaining Science* [1988]). As a first step, we describe and explore a *model system*, and then we compare it to a *target system* we are trying to understand. Giere's main examples were ideal

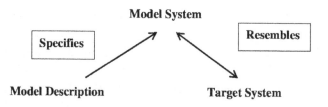

Figure 13.1. How models relate to their targets

springs and pendulums, as analyzed in physics. There are examples in many sciences—ecologies with only two or three species interacting, perfectly rational agents within economic markets, neural networks where the neurons are influenced only by a few synaptic connections. This is *model-building*, or *modeling*. The word "model" has a lot of uses in science and in philosophy. The word is sometimes used just to mean *theory*, or *theory I am not sure about*. Here I am talking about a particular style of scientific work, one that involves a two-step strategy: understand a model system first, then try to compare it to a real-world target.

One might wonder whether all science is like this. I don't think so. Consider Darwin's *Origin of Species*. Here there is no attempt to detour through a simplified system. Instead he just tries to describe what is going on and gives evidence for his view. Michael Weisberg (2007) uses the example of Dmitri Mendeleev, who developed the first version of the periodic table in chemistry, as another non-modeler. What Darwin and Mendeleev did is different from what Maynard Smith and Price did when they developed the hawk-dove model, and different from what a person does when trying to describe a pendulum that is unaffected by friction, and so on.

Deliberate simplification is common in science. Sometimes it seems that we start with the real world and imagine away some complications—this is also called *idealization*. In other cases, it seems that what we are doing is constructing a fictional or hypothetical scenario from scratch, and comparing it to the real world once we understand it.

This idea of comparing imaginary systems to real ones can be philosophically puzzling, and an idea we might try to avoid. One view of scientific modeling is that all the real work is done by mathematics, and the rest is just colorful talk. I presented the hawk-dove model above in words, but it would usually be done mathematically. This move would not solve all the problems, because there is a difficult question of how mathematical structures themselves relate to the natural word. But it might help. Abstract mathematical models might be thought of as attempts to mirror some of the dependence relationships that exist between the parts of a real system. A mathematical model will treat one variable as a function of others, which in turn are functions of others, and so on. In a mathematical model, a network of dependence relationships can be constructed, and the result can be compared to the dependence relationships in a real-world target.

On the other hand, not all models are mathematical. There also seems to be a close analogy between models of the sort that are written down and models of the sort that are actually built. There are carefully built physical scale models of the San Francisco Bay and Mississippi Delta, for example, that are used to predict the consequences of changes that might be made or might naturally happen to these parts of the world. In those cases, it is pretty obvious what people are trying to do. We build a simpler and smaller version of the system we want to understand, study the simpler case, and then export some conclusions back—cautiously, with qualifications—to our dealings with the real system. There is a way of looking at scientific model-building that makes it seem a lot like this. In some cases we build the simpler system; in others, we merely imagine it, but in a rigorous way.

Computer models are an interesting case, and in some ways they seem to lie "between" mathematical or imagined models and physically built ones. People doing computer modeling work often say things like: "The population I am studying is inside a computer." I think that does not make sense. What is really happening? Weisberg, in his book *Simulation and Similarity* (2013), treats computer simulations as mathematical models of a special kind. The aim of this kind of modeling is to specify a set of procedural or algorithmic dependencies, and then claim that these are similar to some relationships that hold in the real world. Another possible view of computer modeling stays closer to the idea that model systems are (or can be) imaginary scenarios. A computer can operate as an aid to the scientific imagination. Our ability to specify interesting systems and arrangements outruns our ability to work out what they would do if they were real. The computer, with its precise and regular operation, helps us with that.

However we handle those issues, *truth* seems to have a different role from usual in this kind of science. A model itself is not true or false. A model is a structure of some sort that can be compared to other things, including the real world. Giere says that when we model something, we also express *theoretical hypotheses*. These are expressed in language and compare the system to the real world: "This model resembles an actual physical pendulum in the following ways. . . ." Those claims can be true or false.

We might also say that *accuracy* is a feature of good models, a feature that is related to truth but a little different. Accuracy comes in degrees. Even this view might not be quite right. Models, whether mathematical or not, have a kind of flexibility of use that is important in scientific work. A variety of people can use the same model while interpreting it differently. One person might use the model as a predictive device, something that gives an output when you plug in specific inputs, without caring how the inner workings of the model relate to those of the target. Another person might treat the same model as a picture of the causal structure inside a system being studied.

Here, finally, is yet another way of thinking about what scientific models of the highly simplified kind do for us. In many cases, these models give us new *conditional* statements, statements of the form "If there is a setup of kind X, then it will do Y." I'll abbreviate this to "If X, then Y." When a model is simplified, or includes other imagined modifications of the world, no actual setups of kind X will exist. The conditional statement is then called a *counterfactual* conditional. These are surrounded by their own philosophical controversies, just as fictions and mathematical claims are. And if the X side of the conditional—the "antecedent"—describes a situation that never actually holds, how can the conditional claim be useful to us, even if it's true? Part of the answer surely lies in the idea of approximation. Often, although there will be no setups that exactly fit the antecedent, there will be some situations that approximate it. Then we might wonder whether we can go from "If X, then Y" to "If approximately X, then approximately Y." This pattern of inference is not a deductively valid one; it can definitely go wrong. But sometimes, with the aid of other information, we might be able to make this move. Then we can learn something of practical importance from the model. Considered empirically, a simplified scientific model can be seen as a device that generates a lot of conditionals, if-then statements, where most of these are counterfactuals, but some are usefully close to nature.

13.5 *Consensus*

I'll now shift gears and turn to a topic that makes connections between themes in several chapters of the book. That topic is *consensus* in science, and the ending of debates.

Consensus is agreement, especially agreement that has been made explicit. In scientific contexts, a consensus usually involves agreement on some particular theory or approach. Sometimes people might agree completely, but there can be consensus without complete agreement and also without complete confidence. In a Bayesian framework, a consensus would be seen when the members of a scientific community have similar, and reasonably high, degrees of belief in some hypothesis.

Disagreement can be healthy, and a lot of philosophers have thought that diversity of belief is a positive thing. Debate is good for science, and it requires that there be different options to discuss—preferably different options that have been worked out in some detail. One view of science emphasizes its creative side above all else. Feyerabend (chapter 6), and to some extent Popper (chapter 4), had attitudes like this.

When thinking about science as the creative exploration of ideas, consensus is entirely optional. How bad would it be, for Popper, if a field was not able to come to a consensus? It doesn't seem to be something he would worry about much. Feyerabend would be positively pleased by this situation; he recommended that scientists continually come up with new ideas and hang onto them, despite their problems, to see how they might develop. A situation of permanent diversity is something he would welcome.

Sometimes, however, there is a need to settle a scientific question, as far as possible, and that applies especially when we have to *do* something in a way guided by science. Many situations involve the export of information from a scientific community to a governmental body, or similar group, that has to make a policy decision. Then a question is often asked: is there a scientific consensus on this question? (On human-caused climate change, for example, or the vaccination of children, or what constitutes a healthy diet.)

In some cases there will be complete agreement within a scientific community, and then things are easy. But in other cases there may be a view endorsed by a majority, but with some dissenters. Then things get more difficult.

One might initially think that they need not get so much more difficult. We looked earlier at an expected value model of action that is comfortable with uncertainty about the way the world is. The best action can be chosen by looking at all the ways the world might be, how probable we think they are, and the consequences of all the available actions in these different states of the world (section 12.5). But to use this approach, we need to decide which possibilities are more probable and which are less probable, and this is something people will disagree about if there is no consensus.

If there is very wide disagreement among qualified people, the problem cannot be readily solved, and certainly a philosophical theory can't help much. There are some sensible things that can be done—we might work out what the really disastrous outcomes are (the disastrous combinations of a policy and a way the world might be) and avoid them. There may also be other good rules of thumb. What I will do here, though, is explore some particular kinds of cases.

In some situations a scientific community might be asked whether some question is settled, and it is found that everyone agrees that the debate is over and also agrees which view has won. I will call that a *spontaneous consensus*. In other cases, there needs to be a more organized attempt to pull things together, and to work out whether lingering uncertainties are serious or not. People have to debate what a public statement should look like. If agreement is then reached, I will call this a *curated consensus*.

"Curated" sounds suspicious, and I do think this process has risks. But I don't want to suggest this is always a bad thing. The risks in these situations tend to be found on many sides, and not easily resolved. A closing off of debate could come too early, or too late. If it comes too early, it can shut down debate that should have continued, and a policy decision using that science might be based on an erroneous view. If it comes too late, the result is wasted time and money, lost opportunities to move ahead, and perhaps different kinds of bad policy decisions. Neither risk is intrinsically more acute than the other.

In an earlier time, when policy makers wanted scientific input, much would have been resolved behind closed doors. That is less and less viable in modern democratic societies with free flows of information. Often now, the situation will require some public recognition of the basis for a policy being applied, so the scientific community may be asked to say something as definite as possible. And while it might be possible in some cases for a consensus to be declared while scientists still work quietly on unorthodox options, public accountability will again make that difficult. If a question is said to be settled, money is not well spent looking at alternative views.

Earlier in this book I distinguished three levels or scales at which we can look at science: (1) the level of the individual, (2) the level of the scientific community, and (3) the level of a larger society that includes a scientific community. Spontaneous consensus is a level 2 matter. People in a scientific community just agree, and can report this if asked. A curated consensus is something that also involves level 3, an interaction between a scientific community and the rest of society or their representatives. There can be feedback between the external consumers of scientific information and a scientific community. In these situations there is a trade-off: we don't want the mere existence of some questioning in the field, perhaps by a few obsessives, to override a sensible use of hard-won knowledge. But we don't want to draw definite conclusions too early.

A case that is worth thinking about is HIV and AIDS. The early 1980s saw the start of the AIDS epidemic (not yet with that name) as a visible problem, affecting especially gay men, hemophiliacs, and intravenous drug users. In 1983, the labs of Robert Gallo in the United States and Luc Montagnier in France announced that a particular virus appeared to be associated with the disease. (A priority dispute followed, as discussed in chapter 7.) The 1980s saw a steady growth in the evidence that the retrovirus HIV was the cause of AIDS. There were some dissenters, notably Peter Duesberg, a University of California biologist who worked on cancer and viruses. In a detailed 1987 article, he suggested that HIV was not a likely cause of AIDS-type ailments, and that the virus was probably a mere "passenger."

In 1988 the U.S. National Academy of Sciences put together a statement summarizing the situation, and said "the evidence that HIV causes AIDS is scientifically conclusive" (website [6]). In retrospect, this seems an early stage to be making a claim so definite. The evidence the National Academy relied on in 1988 was entirely based on observed associations, as they acknowledged. People sometimes say that it's not possible to infer causal relations from observed associations. I think that is overstating the problem; associations can make the existence of a causal relationship very likely (see chapter 11). But experiment certainly makes a big difference in establishing causal claims. The mechanisms by which HIV might do so much harm were also unclear at the time.

I was living in California then and watched as discussion of this question developed. I knew a doctoral student in oncology at the University of California, San Diego, where I was studying, who was concerned about Duesberg's arguments and took them seriously. He regarded Duesberg as a leading authority on viruses of this kind, and found it hard to simply write off his dissent.

What did Duesberg think the real cause was? He argued that taking recreational drugs, other lifestyle issues, and AZT itself (the main early treatment of AIDS) stressed the immune system and might cause AIDS-like symptoms. A problem for his view was the fact that hemophiliacs, who need regular blood transfusions, had very high rates of AIDS before blood screening came in, and they did not have much else in common with other patients.

AIDS became a major problem in sub-Saharan Africa around this time. The government of President Thabo Mbeki in South Africa became interested in unorthodox views of the epidemic. In 2000 Duesberg, along with other skeptics about HIV, served on an advisory panel to Mbeki. The Mbeki government decided not to approach the epidemic with the antiretroviral drugs used elsewhere. In reply to the storm of criticism that resulted, Mbeki sometimes said that he was interested in an open-minded questioning of theories and the exploration of diverse possibilities. Surely that is a properly scientific attitude? In some ways it is, but the need to make a good policy decision was the other side of the situation. This period lasted roughly from 2000 to 2003, when the South African government changed course. Analyses done later concluded that several hundred thousand people died as a result of the Mbeki government's policy in those years (Nattrass 2008).

There is a quote from Albert Einstein that Duesberg has used: "The important thing is to not stop questioning." (Unlike some Einstein "quotes," this one he probably did say.) Nothing is certain, and history has shown that big surprises do come. That is true, but we also have to act, make choices. The evidence that HIV was the cause of AIDS was strong and accumulated steadily through the 1980s and 1990s. The question of treatment was a partially separate issue; AZT is a very toxic drug, and it was a hard decision for many to take it even if mainstream theories were accepted (a theme portrayed in the 2013 movie *Dallas Buyers Club*). There was widespread suspicion, not unreasonably, of drug companies. That was an incentive to look at other options.

The mainstream approach was vindicated, however, and in time the details of how HIV leads to illness were worked out. Better drug regimes were developed in the mid-1990s, "combination therapies" that did more good and less harm.

What is the message of this case? How might it inform other cases that are ongoing? I am not sure. The official 1988 statement about HIV and AIDS, an attempt at curated consensus, makes me uneasy given its timing, but its main claim was true. The case also reminds us that there is a downside to the initially appealing idea that it's always good to be as open-minded as possible and avoid closing off debate. You can keep questioning, but often you also have to act.

It would fine to keep questioning if you could have policy guided by the weight of expert opinion (the "center of gravity" in the scientific community), without that opinion being undermined by doubters. That is the best-case scenario. But this has become harder and harder to achieve in recent cases, especially in the important case of human-caused climate change.

13.6 *Occam's Razor*

In the discussion of modeling earlier in this chapter, a conspicuous role was played by simplicity. Models are often simplified versions of a real-world system, and we can deliberately simplify in order to understand. That is one role for simplicity in science. In this section we'll look at another one.

Many people think that if we are choosing between rival theories, simple theories are preferable to complex ones. "Occam's razor" is a rule that is usually expressed something like this: *do not postulate entities beyond necessity.* We should shave off excessive complexity whenever we can. "Parsimony" is another term for simplicity, and Occam's razor is sometimes also called the principle of parsimony.

William of Occam was a fourteenth-century English theologian. Apparently it is not clear whether he should be so closely associated with the principle, but he is. Many leading scientists have expressed views of this kind. Isaac Newton, in his *Principia* ([1687] 1999), offered explicit rules for doing science, and "Rule I" was: "We are to admit no more causes of natural things than such as are both true and sufficient to explain their appearances." When Newton refers here to a "true" cause, he does not mean the cause actually working in this particular case, but a cause of a sort that we have learned is genuine and can operate in cases like this (or something similar—there is debate over exactly what Newton meant). Philosophical accounts of knowledge often include a central role for simplicity. Quine's "web of belief" view, discussed in chapter 2 and elsewhere, is an example. Quine saw the totality of our belief system as a "man-made fabric which impinges on experience only along the edges." We have to keep the "edge" of our belief system compatible with experience as it comes in. Every other adjustment we make to the system "has as its objective the simplicity of laws" (Quine 1951a, 39, 42; "objective" here just means goal).

What exactly is the simplicity we are supposed to seek? First, fewer or more entities might be used in a theory. Alternatively, perhaps what matters is not fewer entities, but fewer *kinds* of things; perhaps it does not matter much if you add more planets to your picture, if you already have planets, but you might want to do as much as possible with planets before you add something else. Third, we might prefer less complicated over more complicated causal arrangements. This can conflict with other kinds of simplicity; you might be able to tell a story with a small number of causal factors, but only if they are arranged in a very elaborate way. Fourth, a simpler theory might be more compact when written down, though this will depend on what language is chosen, and it's hard to see it as an important matter except as a guide to other kinds of simplicity.

Simplicity has been valued for a number of reasons. First, it might have an *epistemic* role; perhaps we should prefer simpler theories because they are more likely to be true. This idea might be hedged by saying ". . . other things being equal" (*ceteris paribus*). It is also possible to value simplicity for other reasons, perhaps because simple theories are easier to work with. This is sometimes called a *methodological* rather than an epistemic preference for simplicity.

There is some artificiality in how this problem is usually set up. People often say that when there is more than one theory that accommodates the data, you should choose the simpler one. But there is more to situations involving evidence than merely *accommodating*—not being knocked out by—the data. As the Bayesian framework makes clear, there can be all sorts of differences in the amount of support that a theory gets from a piece of data. In addition, often there is no need to choose between the theories at all. One can keep an eye on several possibilities, maintaining different degrees of confidence in them, including some simple theories and some that are more complicated.

Moving past these issues, what can be said in favor of a preference for simplicity? I think sometimes people think this preference *must* be justified because it would be a disaster if we had to give it up. But that is not much of a justification. If, on the other hand, simplicity is preferred merely on grounds of convenience, that seems innocuous, but it would be reasonable to ask why such a person is not trying to get to the truth. We might also deliberately simplify in model-building (section 13.4). But this is a different sort of case; it is not an application of Occam's razor if we know that a simple model is inaccurate in some ways but we think it represents some aspects of a system in a way we can understand. The more important issue is whether it is justified to prefer simple theories because this is more likely to get us to the truth. If so, why should this be? Why should the world be simple? If God made the world, then we might expect simplicity because it is more apt or fitting. Seventeenth-century versions of a simplicity preference were often based on this assumption. That sort of thinking still seems to stay around—we are sometimes told that Nature (with a capital "N") does not do things in vain. But once God is out of the picture, this is baseless.

Simple theories that also work well do have a kind of pleasing character—getting a lot from a little is pleasing. But the fact that it *would* be good if a simple theory worked well does not tell us that such a theory *will* work well.

The history of science is also sometimes said to favor a simplicity preference, but I don't see much evidence for this. To pick one example, Michael Faraday, the experimental pioneer of electricity and magnetism in the nineteenth century, apparently doubted Dalton's theory of chemical elements because it seemed too complex, with dozens of elements (in its later versions), compared to other views at the time that had far fewer basic entities. This was not a very good call, and Dalton's view was a big step forward. In chemistry, the simplicity of pictures of the world has gone up and down. In ancient views there were often four or five elements, then with Dalton and Mendeleev in the nineteenth century there were dozens, growing to over a hundred. Bringing in physics, it was later possible to explain all those elements in terms of three subatomic particles: the electron, proton, and neutron. Then those particles got broken down and more particles were added, reaching about thirty today. Simplicity comes and goes. Who knows where things will end up?

I'll now give my own view, one that gives a much smaller role to simplicity than most people accept.

First, there are some special cases—areas in science where simplicity of a particular kind is preferable in hypotheses because of how we think that part of the world works (Sober 1994). In biological evolution, for example, favorable mutations are rare events. Most mutations either do nothing or are bad for the organism. So evolutionary hypotheses that require smaller numbers of favorable mutations are generally preferable to those that require many. They require fewer unlikely events to happen.

The next idea takes more spelling-out. It starts with an idea from Popper (chapter 4). Popper thought that simple theories, in many cases, can be easily falsified if they are false. So it is good to work with them; they take risks in the way he liked. There is no reason to think that a simple theory is true, but it is more easily shown false if it is false, and that is a virtue.

An example used by Popper and others is mathematical. Suppose we are getting data in the form of (x, y) coordinates. It takes fewer data points to falsify the hypothesis that the underlying relationship has the form of a straight line than it does to falsify the hypothesis that the function is a quadratic, or one with still higher powers of x. The straight-line hypothesis is simpler, it seems, and the simpler theory also takes more risks. I think this case is not so clear, and will take a closer look at it in the notes at the end of this chapter. But for the moment, let's grant the main idea: simple theories sometimes take more risks, and that makes them easier to falsify if they are false. When it holds, this relationship also has a flip side. Think of the situation within a Bayesian framework. If a hypothesis forbids a lot of possible observations—if it rules a lot out—then not only is it easily falsified if it is false, but it also tends to get a lot of support if the observations do turn out to be compatible with it. If a hypothesis takes risks and survives, then its probability tends to go up a lot. I am hedging what I say here, because as always in a Bayesian context, things depend on what the other rival hypotheses say, and other details. But if a hypothesis gives no likelihood, or very little, to nearly all the possible observations, then it will have to give a lot of likelihood to the few remaining. In contrast, a hypothesis that is friendly to lots of possible observations can't be *that* friendly to any of them.

Here is a simple example. Suppose you are doing an ecological study and want to know how many species of bat there are in a forest. For some reason, you think there is either only one species, S_1, or two species, S_1 and S_2, which would be equally common. Suppose that you also assign these two hypotheses equal prior probabilities. You then go out and start looking for bats, making sure you have the same chance of encountering all the individual bats in the forest. The first three bats you see are of species S_1. The hypothesis that the forest only contains S_1 took a big risk—a single bat of another kind would knock it out—but survived. The other hypothesis also survives, but the successful piece of risk-taking by the S_1-only hypothesis will show up in its posterior probability. This probability goes to nearly 0.89, and the probability of the other hypothesis drops very low.

This case illustrates the way that risk-taking hypotheses, the ones that "constrain the data" a lot, get a boost when things do turn out the way they require. Is it generally true that simpler hypotheses constrain the data more than complicated ones? Sometimes I think they do. In the bat case, simplicity did seem to have this effect. However, background assumptions that could be added to a case like this might turn the one-species hypothesis into the less simple one. Suppose we think it is usually true that forests contain multiple bat species, so a one-species forest would be an outlier, an anomaly, something that would clutter up our overall view of bat ecology. It would still be a hypothesis that constrained the data a lot and got a big boost when the data came out in its favor. I think simplicity assessments are often a bit "soft" in these cases, and this is a problem with taking Occam's razor too seriously. Occasionally people say that the defining feature of a simpler hypothesis is the fact that it constrains observations more than other hypotheses—this is what simplicity *is*. I think there is no reason to believe that, other than a desire to get as simple (!) as possible a relationship between simplicity and testing.

The phenomenon described here—the way that simple theories can constrain data more, and hence get more support when things go well—is sometimes called "Bayesian Occam's razor," or the "automatic Occam's razor." There is a bit of an irony here, as the point is well approached by starting with Popper and his liking for risk-taking, but what is being described here is an effect involving confirmation or positive support, which Popper did not believe in. Importantly, Bayesian Occam's razor is not an extra principle, a maneuver, something we can wheel in to deal with problems. That is how many people have thought about Occam's razor. Instead it is just a general fact about the Bayesian framework: given that it rewards risk-taking in a certain way, it can also sometimes reward simplicity.

I'll add one more point about simplicity. I see this as a more informal motivation for a simplicity preference that has some of the same flavor as Popper's idea. If you start with a simple theory, one with only a few causal factors, then when observations come in that clash with your view, you might be able to make more sense of what is going on than you could if you were working with a very complex theory. So it can be reasonable to start out working with simple views even though you expect to get pushed toward more complex ones. The "pushing" works better—more clearly—when you do this.

The leading twentieth-century evolutionary biologist John Maynard Smith had an attitude of this kind. He sometimes defended using a picture of evolution that is notably simple, as it assumes that mutation and natural selection can generally find near-perfect solutions to the problems that organisms confront. This view of evolution ignores a lot of factors, such as constraints deriving from the developmental sequence of an organism, and others. Maynard Smith did not think that evolution *is* that simple, but he thought it works well, when thinking about a particular case, to start with a simple framework and use it as the basis for thinking about the more complicated possibilities.

This sort of thinking is related to the use of deliberately simplified models, discussed earlier in this chapter. In that other style of work, we deliberately simplify in order to achieve understanding of some relationships at work in a complicated system. Here, we don't impose simplicity but use it as a starting point. The two approaches are fairly similar; a person might say, "Let's first assume this system is simple, and see how we go," or, "I know this system is not simple, but I will pretend it is, because that will help me understand some of its features."

Earlier in this section I distinguished methodological and epistemic versions of Occam's razor. The account I have been sketching over the last page or so is perhaps a mix of the two. I initially summarized the epistemic version of Occam's razor like this: "We should prefer simpler theories because they are more likely to be true." That, I think, is not right at all. But compare this: "In many situations it is good to start by working with simple theories, and see how they fare, because this is more likely to get us to the truth in the end." Something like that might be right.

Further Reading and Notes

Simon Blackburn's *On Truth* (2018) is a good introduction to the problem. Other relevant treatments, very different from each other, include Horwich (1990), Devitt (1997), and Price (2003). My use of the term "satisfaction condition" in section 13.3 is a bit unorthodox, as it is more common to use this term for a feature of predicates (like "red" and "electron") than to apply it to sentences, or to use it for the conditions of satisfaction of a desire, rather than a belief or claim about how things are.

For a general discussion of theories of representation in science, see Frigg and Nguyen (2016). There is a large literature on models and model-building. In addition to the works in the main text, see Levy and Godfrey-Smith (2019), a collection on the imagination; Sterrett (2002), on scale models; Frigg (2010), Thomson-Jones (2010), and Godfrey-Smith (2009), on fictions; Levy (2015), on doing without fictional systems; and Suárez (2009), another alternative view. Godfrey-Smith (2019) discusses the connection to conditionals. Downes (2020) is a concise introduction to many issues around modeling. Potochnik (2017) is a detailed treatment of idealization.

Oppenheimer et al. (2019) describe the formation of consensus in various parts of climate science and argue that risks were sometimes understated in this field because of a wariness about sounding alarmist. Between the writing of this book and its production stages, the COVID-19 pandemic of 2020 appeared. The situation is evolving quickly as I write, and its messages are not yet clear. But the pandemic raises questions in many areas covered by this book, especially around the formation of consensus, interactions between scientific work and practical projects, and the use of idealized models as means for understanding and prediction. There will be much material here for philosophy of science in the years to come.

On simplicity and Occam, Baker (2016) is detailed and helpful. Sober's 1994 paper "Let's Razor Ockham's Razor" influenced my thinking. Sober's more recent treatment is *Ockham's Razors: A User's Manual* (2015); as the two titles suggest, he now takes a different view. Kelly (2004) defends, in much more detail, a view of simplicity related to the position sketched at the end of the chapter.

Here is a follow-up of a point I introduced briefly in the section on simplicity, which has to do with the relationship between simplicity and falsification. Initially, it seems clear that the hypothesis that a function relating two variables is a straight line is easier to falsify (in principle) than the hypothesis that the function is a quadratic ($y = ax^2 + bx + c$), or one with higher powers of x. Three data points can suffice to show that a curve cannot be any straight line, but four are needed to show that the curve is not a quadratic, and so on up. The hypothesis that the curve is a straight line seems simpler, too. However, this is a situation in which the simple-looking hypothesis (straight line) is a special case of the more complex one (quadratic), as a straight line is a quadratic with the parameter a equal to zero. So really, the hypothesis that the function is a quadratic is a *broader*, or less specific, hypothesis, one that allows both for options that look simple (straight lines) and others. The broader hypothesis is also harder to falsify. People sometimes argue that the breadth of the quadratic hypothesis is a kind of complexity, and hence one that does relate to falsification. I think that in a case like this, it is the shapes of the functions themselves that are simple or complex, although as seen also in the examples in the text, it can be unclear what simplicity amounts to. (Is a sine wave complicated because the curve changes direction so often, or simple because it follows such a regular rule?)

Chapter 14

The Future

14.1 *Empiricism, Naturalism, and Scientific Realism*

We now reach the end of this hundred-year tale. This final chapter has two sections. The first looks back over some of the main themes of the book and tries to put some of the pieces together. I'll single out some ideas I think of as being on the right track and contrast these with others I see as mistakes. The last section will look at some changes in science over time—not changes in the content of science, but changes to its organization—especially changes that are in process now, which the ideas in this book might help us think about.

I'll initially organize this looking-back section by returning to three named philosophical theories or projects, the three big "isms" that have had some endorsement in this book: empiricism, naturalism, and scientific realism. In the first edition of this book, the last chapter included a deliberate attempt to put those three together, and I endorsed the resulting combination. This time I have a few more reservations about each of them. The discussion here might still be seen as fitting the three together; I don't think they grate against each other as much as some people suppose. After an initial recap, focusing especially on empiricism, I will proceed by drawing ideas from each, putting them together into an overall view.

Empiricism is where the story started. We looked at this general approach to understanding knowledge, and at an ambitious version of empiricism that was developed in the early twentieth century. That view and its problems set the events of many later chapters in motion. What should we make of empiricism now?

I will start by looking at a couple of features of the tradition that I see as almost entirely mistaken.

The first idea is the approach to evidence that many empiricists have taken, an approach that focuses on the confirmation of generalizations by means of observation of particular cases or events that fit, or fail to fit, the generalization. Even when it was agreed that it can be only part of the story, this picture took over the discussion of evidence—not entirely, but in large part—for many years. Looking at it diagnostically, this approach might be seen as a consequence of the coming together of two themes, one old and one new. The old one was the traditional empiricist tendency to think about experience in terms of a succession of individual impressions or sensations, one after another, like mental or experiential atoms. The new theme was the development, around the turn of the twentieth century, of mathematical and logical tools that promised a more sophisticated treatment of what a succession of sensations (or observations, more generally) might tell us.

Once you start to think about things in that way, it can become very appealing as a framework. It generates a kind of model-building exercise, where one tries to do as much as possible with these slim resources. The project is appealing because it sets things up in a clear way, and because the problem of evidence seemed such a mess outside this model. We did learn some things from attempts to carry out this project, but much of what we learned was negative; we learned a lot about what will not work. Here I have in mind all the problems covered in chapter 3. And even the activities in science that are closest to that traditional empiricist picture—closest to a situation where we collect particular cases and try to generalize from them—have a crucial difference from what the empiricists envisaged. This I argued in section 12.6, where I looked at the primacy of procedures in understanding evidence of this kind.

Related to this view of evidence is another large-scale mistake in traditional and much modern empiricism. This is the attempt to restrict what we can know about and investigate to patterns in experience, or patterns in the observable realm. I accept that sometimes in science it has been fruitful to pay very close attention to exactly what the data are, and to what we can observe and how observation works. Einstein is the great example here, influenced by Mach and inspiring empiricists of his time in turn (though Einstein's commitment to strongly empiricist ideas is sometimes overstated; he was not a phenomenalist and did want to work out what lay behind appearances). It would be foolish to deny the importance of an episode like this, but as a general view of science there is little to say for the idea that it's merely about what we can observe.

As noted in the first chapter, empiricists, as well as some others, have often operated with a picture of the mind's access to the world that has been called the "veil of ideas." The mind is seen as confined to its own sensations and thoughts, trying in vain to reach a hypothetical world beyond. Many philosophers now agree that this is a misleading picture, but it is easy to fall back into relatives of this view, both when thinking generally about experience and also when thinking about science. Philosophy of science has often hung onto enough of the old picture for trouble to arise. It is easy to fall back into a framework in which we distinguish two layers, or domains, in the world. One domain is accessible to us and familiar—the domain of experiences, or the domain of the observable. The other is inaccessible, mysterious, "theoretical," and problematic.

Another theme in the book concerning observation has been the relation between observations deriving from experiments as opposed to more "passive" or happenstance kinds of observation. At the end of chapter 12 I brought together a couple of lessons here. One involved, again, the importance of procedures in turning events into evidence, and the other involved the role of intervention when learning about causal relationships. For many years empiricism paid little attention to the special features of experiment. Some of this might be due to an overall neglect of action in the empiricist tradition, a neglect that extends through to more modern empiricisms such as the view of Quine.

Many empiricist positions have also neglected the social side of knowledge and science. There have been exceptions, in unorthodox thinkers such as Peirce (1877) and Dewey (1929a) and some recent views. But this has been another case of neglect that, in retrospect, looks quite unwise. It's not merely neglect of something that might be seen as worth discussing; it leads to mistakes. We can see Kuhn, for example, as arguing that science cannot be adequately described with anything like the concepts seen in mainstream empiricism, because science is a much more complicated machine than anything those concepts are able to represent.

Even setting aside special features of science, from the point of view of its opponents, empiricism is based on a hopelessly simple picture of what knowledge involves. Empiricism is often summarized by saying that the only source of knowledge is experience. But what is this talk of "sources" doing? We ask, is there just one source of knowledge, or more than one? This is like asking, is there just one pipe leading into this tank, or more than one? But learning about the world is not a matter of taking in a substance through a channel; epistemology is not plumbing.

Does all this destroy empiricism in the philosophy of science, or motivate us to reform it? One response at this point is to say that there is nothing to be gained by aligning oneself with the empiricist tradition today; it should just become a part of history. Another response is to urge that we not miss the forest for the trees. There is still something big that empiricism gets right.

I find myself closer to the second response, but with reservations. And when I think about why empiricism still seems important, what recommends it is not so much a general philosophical theory, but an outlook or mindset, one that can be seen in science, philosophy, and elsewhere. This mindset features an ideal of open-mindedness and intellectual flexibility, a distrust of dogmatism, and a grounded, rather than other-worldly, orientation.

If we value those things, though, we are valuing things that can be chosen, not things dictated to us by the general nature of language or the principles on which human minds work. What then remains of the idea of empiricism as a general philosophical theory, an account of knowledge in general?

There does seem to be a tension there, but empiricism has for a long time tried to be a combination of general theory and recommended mindset. It has often been a story about how the mind (or language) works, along with a recommendation to do things in a way that best fit with this picture—not to deceive oneself about possible sources of knowledge, to be mindful of our fallibility, and to retain openness and flexibility. We might see empiricism as a general theory (a philosophical theory) of what is going on in the more successful strands in our multifarious attempts to learn about the world.

Leaving this theme for a moment, let's turn to a second "ism" that was partially endorsed in the book, scientific realism. One aspect of the turning away from very traditional forms of empiricism has been a move toward scientific realism, which includes a denial of the view that all we can learn about is patterns in experience, or patterns in the observable domain. It is quite hard to come up with a good statement of what scientific realism is. But there is a general picture of people, including scientists, and their place in things that is characteristic of realism in general. This picture sees us all as parts of the same complex, structured world, learning about that world when things go well, and acting on this larger world in constrained but partially efficacious ways. I think that this outlook is right, and attempts to motivate a move away from it, including attempts discussed in chapter 10 of this book, do not work.

A third view, endorsed in chapter 9, was *naturalism*, which I regard as an orientation to philosophy rather than a view about what the world contains. Naturalism of this kind involves being willing to draw on our scientific picture when working out how science and other kinds of knowledge work. The side of naturalism that I don't accept is the side that aims to constrain philosophical exploration. A particular kind of freedom is important in philosophy; you never know where it is going to go next, and where the next interesting idea might come from. When naturalism falls into a constraining mode (do things *this* way and *don't* do that . . .), I step away from it.

Now I will put some of those ideas, the good ones, into a sketch of an overall view, doing so without worrying too much about the labels, the "isms."

Humans are biological organisms embedded in a physical world that we evolved to deal with. All our lives—including the most elaborate outgrowths of our social and intellectual lives—involve constant causal traffic and interaction with a larger world in which we are embedded. Our attempt to know about the world is one aspect of our causal interaction with it; much of this interaction is more practical. Our perceptual mechanisms—eyes, ears, and so forth—are tools that we use to coordinate our various dealings with the world. These mechanisms respond to physical stimuli caused by objects and events in our environments. From the inside, we can never establish with complete certainty what lies behind a particular sensory input. But looking at ourselves from "sideways on," from the point of view taken by biology and psychology, we can establish regular principles concerning how our perceptual machinery responds to objects and events distant from us. We can work out how our perceptual machinery helps us navigate the world.

So far this is not a description of science, or even human beings in particular. It is a claim about all the organisms that use perceptual mechanisms to adapt themselves to what goes on in their environments. But this point is enough to help us avoid some philosophical problems about our "access" to the world. We should not think in terms of two domains in reality, one accessible and one mysterious. We are biological systems embedded in a world containing objects of all sizes and at all different kinds of distance and remove from us. Our mechanisms of perception and action give us a variety of different kinds of contact with these objects. Our access to the world via thought, theory, observation, and experiment is a complicated kind of causal interaction. This access expands as our technology improves. Parts of the world that must, at one time, be the subject of indirect and speculative inferences can later be much more directly observed, scanned, or assayed.

Though this situation is common to all humans, as a consequence of their shared biological nature, there are differences in how different individuals and intellectual cultures approach the problem of investigating and understanding the world. One important point of disagreement is in how people handle the assessment of big ideas—big theories and explanatory hypotheses about the world—as opposed to how they handle everyday life. Maybe our shared biology is enough to make us all fairly empirical when we are trying to get food to eat and work out how to get home. But this does not apply to attempts to develop and justify theories about our overall place in the universe. Here we find sharp disagreements in approach.

Science is in part a strategy chosen in response to this. The scientific strategy is to construct ideas, and embed them in surrounding frameworks, in such a way that exposure to experience is sought even in the case of the most general and ambitious hypotheses about the world. This strategy involves a kind of dialogue between an imaginative voice and a critical one, where the critical one looks to experience as a constraint on hypotheses. In science, an acceptance of the *provisional* status of big ideas is at least part of the ideology—part of the official mindset that one is supposed to have—even if people don't always live up to it. In other forms of intellectual culture, especially where tradition has a stronger role, this is not so.

We can distinguish this general scientific strategy from a particular way of organizing how the strategy is carried out. In a broad sense, science is the attempt to assess large-scale ideas by exposing them to experience. The Scientific Revolution and the work that followed it developed a particular, socially organized way of carrying out the strategy. The term "science" can also be used, more narrowly, to refer to that social organization.

An individual, all alone, could set up a self-contained program of formulating hypotheses and assessing them via observational testing. An individual can internalize science's dialogue between the imaginative and critical voices. Such an individual might refuse to trust others, and attempt to get as close as possible to the old fantasy of the lone empiricist relying entirely on his own experience. That is a possible way of carrying out the scientific strategy, but it is obviously far from the usual ways and the most successful ways. Through the development of institutional structures, reward systems, and informal norms, a particular way of socially organizing empirical investigation has developed.

The distinctive features of science as a social structure are found along two dimensions. One has to do with the organization of work at a given time. This is where we find the reward structure, the mix of competition and cooperation, norms of criticism, and the like. The other dimension has to do with the relationships between different times, and with the transmission of ideas between scientific generations. Scientific work is cumulative. Each generation builds on the work of its predecessors. This requires both trustworthy ways of transmitting ideas across time and (again) a reward system that makes it worthwhile to carry on where earlier workers left off.

With a social structure of this kind, the "dialogue between imaginative and critical voices" can become a genuine to-and-fro. We have social mechanisms in place that reliably bring about the checking and scrutinizing of ideas. Some rather dogmatic individuals can work within the system and perhaps play a useful role, provided that flexibility and open-mindedness is found in the community as a whole.

Once we distinguish the two dimensions of science's social organization—the organization of work *at* a time and the organization of work *across* times—some historical questions arise. Were there relatively sudden transitions that gave us these features, or did they evolve more gradually? Did they arise together or separately? The cumulative structure of scientific work is something that is old in some fields and newer in others, and it can be gained and lost in a partial way. As Toulmin and Goodfield (1962) note in their historical account of the development of ideas about the universe from the ancient period to the early part of the modern, a cumulative structure sometimes arose and then faded. A sustained line of work would be set up by a "school," often in some particular city, and then it would recede and be replaced by a succession of individuals working alone, "reinventing the wheel" over and over again. Gradually, field by field, this haphazard pattern was replaced by more cumulative work.

Turning to the organization of scientific work *at* a time, the middle of the seventeenth century may be crucially important. As Shapin and Schaffer (1985) emphasize, this period saw work directed at setting up a culture of controlled criticism and a new kind of network of trust. This side of science's social organization was deliberately shaped, rather than arising by happenstance.

The case of alchemy is interesting here. Alchemy was the precursor to chemistry, and it was influential through the end of the seventeenth century. (Newton was quite invested in it.) Alchemy was a combination of practical work based on detailed recipes and esoteric, speculative theories. (Chemical reactions were signified in astrological relationships between planets, for example). Alchemy was quite empirical in some ways—very results oriented—but the work of alchemists was organized in a way that contrasts strikingly with modern science. Alchemy was often intensely secretive; rather than accessible dissemination of results, there was a culture of private and restricted communication. This was partly because of the mystical side of the field, and partly because of the hope for massive financial benefits via finding a way to transmute other metals into gold. As Shapin and Schaffer emphasize, Robert Boyle contrasted his open, cooperative new scientific culture with the secrecy of the alchemists, as well as with the emptiness and dogmatism of Scholasticism.

These social accounts of science have considerable power, especially when they are used to highlight the weaknesses in traditional empiricist ideas. But the social accounts, when they get traction on the phenomenon, tend to assume some basic empiricist ideas about the constraint of hypotheses by observation as a background feature. (We saw this in Hull's view, and it was a point of contrast with Latour's, in section 9.4.) A willingness to be guided by data, rather than merely using data as an argumentative resource, is part of the practice.

Once at the end of a course I taught, a student asked a question looking for an overall message from all the details we'd covered. She asked: do I think scientists are mostly unbiased, or is there lots of bias in science? In reply: scientists are people, and people are often full of biases. As it happens, among the scientists I know, quite a few are notably fair-minded as individuals. But individual virtues, or the lack of them, are not the main issue here. The more important thing is the extent to which the norms and habits of scientific behavior give the process a self-correcting character. If bias induces error, there will usually be something to gain from showing it; there is a process by which correction will come.

All of this is very present-tense. Why assume that things will stay the same?

14.2 *Another Kind of Scientific Change*

The hundred years or so we've covered are in some ways *the* hundred years of self-conscious philosophy of science. More accurately, there have perhaps been about 130. In other ways this is not so, as has been obvious from the many excursions this book has taken into earlier centuries, especially the seventeenth and eighteenth. But philosophical attention to science per se became more focused during the period covered in this book.

This was also a particular period in the history of science itself. It is tempting to think of science as a stable achievement, something that originated, developed into a mature form, and now continues on. But science changes in an ongoing way as well, not just in the theories people believe and the technologies they use, but in its organization and patterns of work. These patterns took particular forms during the twentieth century, and the way science is run has changed a bit since then—it has been changing even over the period between editions of this book. This last section will look at some of these changes.

I will start by looking at changes in the relations between scientific communities and the rest of society—some level 3 facts, in the sense introduced in chapter 5—and some related questions about "pure" as opposed to "applied" science.

Kuhn, once again, gives us a way in. He did not have a lot to say about the relations between scientific communities and society at large, as he saw science as a rather insulated and self-governing activity. For Kuhn, change in science is propelled from within. Guidance from outsiders will usually be irrelevant, unless they turn off the money or interfere in a large-scale way. Kuhn was writing in a special context, the period of the Cold War between Western democracies and the Soviet Union. In the United States there was considerable enthusiasm for science, prompted in part by anxiety about falling behind the Soviets. The result was a high level of support and a practice of not worrying too much about how creative people would use that support. Self-propelled normal science was a natural result. Some features that Kuhn saw as characteristic of modern science in general were especially marked at this time.

Scientific communities are now becoming less insulated and less self-directing. Some might say that this aspect of Kuhn's view was exaggerated even for the period he had his eye on, but however true it was then, it is less true now. Contemporary science has more to-and-fro with business and the rest of culture, and more oversight and guidance from governments, funding bodies, and administrators of many kinds. The links between science and business, especially in molecular biology and now in nanotechnology, lead to a lot more money going into the pockets of scientists. Many of them start companies to commercialize their work. In maintaining research, writing grant applications is now continual and onerous; it cannot be assumed that support will continue if a lab does good work. There was always oversight, but a light touch has been replaced with a heavier one.

These changes are having a range of consequences. Many would say first, and with good reason, that the changes have led to more democratization of science, through a weakening of "old boy" networks. Because of more competitive and ubiquitous grant-writing processes, funding has less inertia and more mobility. Those same processes have led to a massive amount of energy being spent just on obtaining grants, though. Some studies have found scientists spending close to 20 percent of their research time working on grant applications (Gross and Bergstrom 2019), rising to as much as half, or more, in the case of medical researchers. Gross and Bergstrom argue that the inefficiencies are so acute that it might be better to give out most funds with a lottery among the basically acceptable projects. Another alternative they raise is to fund researchers based on past success—in practice, this would be a move back toward earlier arrangements where there was more inertia in the system, a shift that would also have a downside.

Another development, and one that I do think it is reasonable to be concerned about, is a weakening of a related feature of Kuhnian normal science. Science has a tradition of rather unconstrained inquiry, of scientists "following their noses" without worrying too much about practical applications for their work or externally set priorities. Kuhn wrote about this as a strength. John Dewey, writing much earlier and in a more broad-brushed way, suggested something similar. He said that part of the transition to modern science in Europe, around the seventeenth century, was the development of a style of inquiry that was new in the way it combined two features that were present but separate in older traditions. Science is an empirical mode of investigation that is not tied down to specific practical projects. Earlier work was either empirical but narrowly constrained (craft traditions), or open and broad but less empirical (classical Greek philosophy and its successors). The early modern period gave us the combination of empirically based work, full of practical tinkering, but led by fundamental questions and not held to a particular set of practical concerns.

This idea in Kuhn and Dewey is not a celebration of the un-useful, a disdain for the practical applications of science. Instead, the picture is that if you let people follow their noses, they will find out things that will eventually be very useful, though we can't predict where they will be found. I said in chapter 5 that I think Kuhn felt a kind of awe at the ability of normal science to home in on topics and phenomena that look insignificant from outside but turn out to have huge importance. This depends on the scientist's interest in internally directed puzzle-solving.

The result is a *practical* justification for "pure" inquiry. There is an illustration of this that is becoming something of a cliché. It is said that Michael Faraday's work on electricity was driven by scientific curiosity, but it eventually gave us the electric light, and "no amount of R&D on the candle" could have done that. This example is not the best, because the electric light was invented earlier by Humphrey Davy, a mentor of Faraday, around 1802, and Davy was more oriented toward practical applications than Faraday. But the underlying thought is a good one. Faraday gave us the electric motor, and no amount of R&D on the treadmill could have done that.

Another example is CRISPR, the gene-editing system I discussed in chapter 7. This came from research into some quirks of the genomes of bacteria, which led to the surprising discovery that bacteria have something like an adaptive immune system that protects them against viruses. About fifteen years into this work, it was realized that the system could be used to manipulate the DNA of other organisms. It now appears that the practical consequences will be very important indeed.

A similar phenomenon has been seen repeatedly in mathematics; parts of the field that seem entirely removed from real-world relevance turn out to have practical importance. G. H. Hardy was an early twentieth-century English mathematician who worked on number theory, with a particular interest in the mysteries of prime numbers. He explicitly said, without a hint of regret, that his work and his whole field were entirely "useless" from a practical point of view (1940). Perhaps the far reaches of number theory were useless then, but they are now central to encryption, which is essential to the Internet economy.

If you take these cases seriously, there is an argument for valuing scientific self-direction. This is not because (or, rather, not only because) pure understanding is good, but because the way to do well on the practical side is to let scientists follow their noses. These features of science are changing to some extent. The model of limited oversight is trusted less than it was, and universities and government bodies engage in closer scrutiny. There is more demand for commercial connections and for work to align with declared priorities. I think this is an unwelcome development; I share some of that Kuhnian awe at the power of self-propelled science.

This does not mean that scientific work should be constrained in all respects. There are areas where I would put the brakes on, far more than has been normal and more than many scientists would like. The one I have in mind is the use of animals in biological and medical research. The problem here is that there is really no limit to the numbers of animals, of many kinds, that would be used in the service of well-funded scientific curiosity. The steady recent advance of knowledge in neuroscience has been accompanied by the suffering of countless legions of mice, and significant numbers of primates and cats. The engine of scientific curiosity has its negative side; it can drive over just about any other consideration. The work done often leads to benefits eventually, and by unpredictable paths, but we need to balance that benefit with what is being done along the way.

I'll discuss one other feature that has been described, and worried about, in recent science. This is the possibility that there is more misconduct, fraud, and careless work than there used to be. It has been suggested that the kind of competition that scientists are now encouraged or required to engage in is one where sheer speed, and also a kind of showiness, count for a lot, and these incentives often now outweigh the risks of being shown to be a substandard worker.

Is outright fraud increasing? Some data suggest so, though it is hard to say. There are more retractions of published articles than there used to be, and the majority of these retractions are now due to misconduct rather than error (Steen 2011; Fang, Steen, and Casadevall 2012). These figures might be the result, at least in part, of more oversight and changes in how retractions are handled (see Fanelli 2013 for a view of this kind, dissenting from the alarm expressed by others). Also, anxieties of this kind are recurring and go far back—a 1977 *New York Times* article reported, based on informal interviews, that rates of fraud were increasing in an alarming way at that time, right in the middle of what a Kuhnian might regard as a golden age.

Along with fraud, there are concerns about a more general "reproducibility crisis" with respect to scientific results, one not due primarily to misconduct, but to carelessness, cutting corners, and, perhaps, overly relaxed statistical standards for the reporting of results in some fields. Here the sheer pace of contemporary science, and the role of high-impact publications in generating prestige and support, may play an important role. Heesen (2018), using a mathematical model in the same style as some of those discussed in chapter 9, argues that failures of reproducibility are an inevitable consequence of the reward structure of science (see also Smaldino and McElreath 2016). In Heesen's model, the benefit to an individual from speedy publication is a key factor. If those benefits were less, other factors (especially risk to reputation) would play a stronger ameliorating effect, from the community's point of view.

Heesen's model is presented as an analysis of *the* scientific reward structure, not how that structure works at a particular time. Reproducibility problems in some fields do go back some distance; it is not all due to very recent science. But I think there is a widespread sense that the problem is worse now, and the increasingly frantic pace of science is a contributor. This is a difficult problem to fix.

These shifts in the practice of science are integrated with a host of broader societal changes. It would be fruitless to try to turn the clock back, even if it were generally desirable. What we need to do is think as intelligently as we can about how science arose, how it gained its strengths and weaknesses, and how we might shape it, as well as utilize it, in a positive way. The philosophy of science is part of this.

Glossary

For definitions and quick discussions of other philosophical terms, I recommend Simon Blackburn's book *The Oxford Dictionary of Philosophy* (2008). Blackburn also gives more detail about many of the terms discussed here.

Sometimes confusion in newcomers to philosophy arises not from technicality, but from slightly different philosophical uses of everyday language. For example, in philosophy the word "strong," when applied to a view or a hypothesis, usually does not mean effective, and it carries no positive (or negative) connotation. "Strong" means something more like extreme, bold, or tendentious. This is related to the use of the term in logic, where a strong claim is one that has a lot of implications. "Weak" in this sense means cautious, hedged, or moderate. Scientists sometimes use "strong" in the same way. So a "strong" version of a view (empiricism, realism, etc.) is not necessarily better than a weak form. Confusingly, philosophers do sometimes talk of a "strong" argument when they mean that the argument is good, or convincing.

After discussing each term below, I indicate the chapters or sections of the book in which the term is important. Terms in boldface have their own entry in the glossary.

There are two terms used in this book that are my own modifications of more standard terms. These are "explanatory inference" and "eliminative inference." In both cases I am avoiding overly broad use of the term "induction."

Abduction. One of the many terms for **explanatory inference**. This one was coined by C. S. Peirce. (3.2)

Analytic/synthetic distinction. Analytic sentences are true or false simply in virtue of the meanings of the terms within them. Synthetic sentences are true or false in virtue of both the meanings of the words and the way the world is. **Logical positivism** treated this distinction as very important, especially because the logical positivists thought that mathematical statements are analytic. Quine argued that the distinction does not exist. (2.3, 2.4, 2.5)

Anomaly. In Kuhn's theory of science, a puzzle that resists solution by the methods of **normal science**. This is close to the word's ordinary meaning (roughly, something out of place). (5.4)

A priori/a posteriori distinction. If something is known (or knowable) a priori, it is known (or knowable) independently of evidence gained via experience. Knowledge that relies on evidence from experience is a posteriori knowledge. (2.3)

Bayesianism. The theory of evidence and testing that gives a central role to Bayes's theorem, which is a provable result in probability theory that describes how new evidence should affect the probability of a hypothesis. Many Bayesians also now accept **subjectivism** about probability. (chapter 12)

Confirmation. A relationship of support between a body of evidence and a hypothesis or theory. Confirmation is not the same as proof; a theory can be highly confirmed and yet be false. **Logical positivism** and **logical empiricism** put much emphasis on the role of this relationship in science, usually trying to analyze it with an "inductive logic." Their attempts were not very successful. (Chapters 3 and 12).

Constructivism (social constructivism, metaphysical constructivism). In the debates discussed in this book, "constructivism" usually refers to a view in which knowledge (and sometimes, reality itself) is seen as *actively created* by human choices and social negotiation. Those who advocate constructivist views often do not distinguish carefully between the view that *theories* (or classifications, or frameworks) are constructed, and the view that the *reality* described by those theories is constructed. I use the term "metaphysical constructivism" for views that explicitly claim that reality is in some sense constructed (10.5).

Correspondence. A correspondence theory of **truth** holds that truth is a relation of correspondence between a statement and the way the world is. Sometimes correspondence is understood in terms of a sort of copying or picturing, but not always. (13.2, 13.3)

Corroboration. Popper used this term for something that a scientific theory acquires when it survives attempts to refute it. Sometimes this looks like another name for **confirmation**, which Popper rejected (4.5). The term is also occasionally used (though not by Popperians) in a way that is roughly synonymous with confirmation or support.

Covering law theory. A theory of scientific explanation developed by the logical empiricists (see **logical empiricism**), especially Carl Hempel. The theory holds that to explain something is to show how to infer it in a good logical argument that includes a statement of a law of nature in the premises. (11.2)

Deductive logic. The well-developed branch of logic dealing with patterns of argument that have the following feature: if the premises of the argument are true, then the conclusion is guaranteed to be true. This feature is called "deductive validity."

Deductive nomological theory (D-N theory). A term sometimes used for the **covering law theory** of explanation, although it only refers to some of the cases covered by that theory, the ones in which the argument used to explain something is a **deductive** argument.

Demarcation problem. Popper's term for the problem of distinguishing scientific theories from nonscientific ones. (4.2, 4.6)

Eliminative inference. A pattern of inference in which a hypothesis is supported by ruling out other alternatives. (It is sometimes called "eliminative induction," even though these arguments can be deductively valid in some cases; see **deductive logic**.) (12.6)

Empiricism. A diverse family of philosophical views, all asserting the fundamental importance of experience in explaining knowledge, justification, and rationality. A slogan used for traditional empiricism in this book is: "Experience is the only source of real knowledge about the world." Not all empiricists would like that slogan. There are also empiricist theories of language, which connect the meanings of words to experience or some kind of observational testing. (chapter 2, chapter 14, 10.4)

Epistemology. The part of philosophy that deals with questions involving the nature of knowledge, the justification of beliefs, and rationality.

Expected value (or Expected utility). To work out the expected value of an action, when you are uncertain about the state of the world, you work out the payoffs (costs or benefits) resulting from that action in each state the world might be in, and then combine those payoffs in a formula along with the probabilities of those states of the world. This gives an overall measure of the value of the action. The formula is given in section 12.5.

Explanandum. In an explanation, whatever is being explained. (chapter 11)

Explanans. In an explanation, whatever is doing the explaining. (chapter 11)

Explanatory inference. An inference from some data to a hypothesis about a structure or process that would explain the data. There are many terms for this idea or ideas like it, including "abductive inference," "inference to the best explanation," "explanatory induction," and "theoretical induction." (3.2, 11.3)

Falsificationism. A view of science developed by Karl Popper. The word "falsificationism" can be used narrowly to refer to Popper's proposal for how to distinguish scientific theories from nonscientific theories (the **demarcation problem**). Falsificationism in this sense says that a theory is scientific if it has the potential to be refuted by some possible observation. The term is also used more broadly for Popper's view that all testing in science has the form of trying to refute theories by observation, and that there is no such thing as the **confirmation** of a theory by its passing observational tests. (chapter 4)

Foundationalism. A term used for theories that approach epistemological problems (see **epistemology**) by trying to show how human knowledge is built on a "foundation" of basic and completely certain beliefs. These might be beliefs about one's own current experiences. (9.3)

Holism. Holist positions can be found in many philosophical debates.

Generally, a holist is someone who thinks that you cannot understand a particular thing without looking at its place in a larger whole. Two kinds of holism are important in this book. *Holism about testing* claims that we cannot test a single hypothesis or sentence in isolation. Instead, we can only test complex networks of claims and assumptions as wholes, because only these whole networks make definite predictions about what we should observe in a given situation. *Meaning holism* claims that the meaning of any word (or other expression) depends on its connections to every other expression in that language. (2.4, 2.5)

Hypothetico-deductivism. A term used both for a method of doing science and for a more abstract view about **confirmation**. The hypothetico-deductive (HD) method is a common description of scientific procedure in science textbooks. The basic steps are as follows: (1) gather some observations, (2) formulate a hypothesis that would account for the observations, (3) deduce some new observational predictions from the hypothesis, and (4) see if those predictions are true. If they are true, go back to step 3. If they are false, regard the hypothesis as falsified and go back to step 2. Some versions omit or alter step 1. Versions also differ on whether the scientist should regard the theory as confirmed if the predictions made by the hypothesis are true. Hypothetico-deductivism is also a view about the nature of confirmation, as opposed to the procedures used in testing. Here, the idea is that a hypothesis is confirmed when it can be used to derive true observational predictions. (3.2, 3.5, 12.5)

Incommensurability. An important concept in Kuhn's and Feyerabend's theories of science. The idea is that different theories or paradigms can be difficult, or impossible, to compare in a properly unbiased way. For example, incommensurability about standards is the idea that different paradigms tend to bring with them slightly different standards for what counts as good evidence or good scientific work. If two paradigms bring different standards with them, which set of standards do we use if we want to choose between the two paradigms? Incommensurability about language holds that key scientific terms (such as "mass," "force," etc.) can have different meanings in different paradigms. So in a sense, people within two different paradigms can be speaking slightly different languages, even if they seem to be using the same words. (5.6)

Induction. A term with a number of meanings. In one sense, the term refers to a method for doing science described in the seventeenth century by Francis Bacon. This method is often described as one in which lots of particular facts should be gathered first, and generalizations and other hypotheses should be based on this stock of facts. In most of the discussions described in this book, this is *not* what "induction" means. Instead, induction is a kind of argument, or pattern of inference, rather than a method or procedure. I use "induction" for inferences in which particular cases are used to support a generalization that goes beyond the cases observed. So these arguments are not **deductively** valid (see **deductive logic**). The logical positivists and logical empiricists tended to use the term more broadly—for any inference that is not deductively valid but where the premises do support the conclusion to some extent. (chapters 3, 4, 12)

Inductive risk. A phrase coined by Carl Hempel to refer to the bad outcomes, especially practical outcomes, associated with a failed induction. (8.6)

Instrumentalism. One kind of opposition to **scientific realism**. It holds that scientific theories can best be seen as tools used to predict observations, rather than as attempts to describe the real but hidden structures in the world that are responsible for the patterns found in observations. (10.4)

Interventionism (or **manipulationism**). An approach to understanding causal relationships, based on the idea that C is a cause of E if the state of C could be manipulated—or intervened upon—with consequences for the state of E. This is usually not seen as an attempt to reduce the idea of causation to entirely non-causal concepts, as "manipulation" is itself a causal concept. Instead, it is a way of understanding causal networks in a rigorous manner. (11.4)

Likelihood. In Bayesianism and in statistics, a technical term referring to the probability that something (e) will be observed, given the truth of some hypothesis (h). So likelihoods are probabilities of the form $P(e|h)$. The term "likely" is often not tied to this technical meaning, though. Sometimes philosophers say "likely" just to mean probable and say "likelihood" just to mean probability. (chapter 12)

Logical empiricism. A term used for the more moderate views about knowledge, language, and science that derived from **logical positivism** and developed after World War II, especially in the United States. The term is sometimes also used for the earlier stage, **logical positivism** (especially by those who think that not that much changed between the earlier and later stages). Logical empiricism was a scientifically oriented version of empiricism that emphasized the tools of formal logic. (2.5)

Logical positivism. A bold and scientifically oriented form of **empiricism** that developed between the two world wars in Vienna, Austria. Sometimes known as **logical empiricism**, though I use that term for a later and more moderate development of the ideas. Leading figures in the movement were Moritz Schlick, Otto Neurath, and Rudolf Carnap. The view was based on developments in logic, philosophy of language, and philosophy of mathematics. The logical positivists dismissed much traditional philosophy as meaningless.

Metaphysics. A term usually now used to refer to a subfield within philosophy that looks at a particular set of questions. These are questions about the nature of reality itself, rather than (for example) how we know about reality. The questions include the nature of causation, the reality of the "external" world, and the relation between mind and body. The term is sometimes seen as referring to any investigation that goes beyond what can be addressed using science. Construed in that way, metaphysics is regarded by many as a mistaken enterprise. (The logical positivists regarded most traditional metaphysical discussion as meaningless.) But in most current discussion, "metaphysics" refers to a set of questions and does not prejudge the right way to address them.

Model. A word with many meanings, leading to frequent confusion. Sometimes "model" is used in science and philosophy of science just to mean a deliberately simplified theory. I generally follow a narrower use of the term (especially in chapter 13) in which a model is a structure of some sort that is used to represent some other "target" system in virtue of a resemblance between them. Models are often, but not always, deliberately kept simple. Model systems may be real-world replicas of a target, mathematical structures, imaginary systems, or computer programs. The relations between these different kinds of model are controversial. The term "model" can also refer to an analogy that is used to *accompany* a scientific theory and make it more comprehensible. The term also has a technical meaning in mathematical logic, where a model is a precise kind of interpretation of a set of sentences, one that treats the sentences as all true.

Naturalism. An approach to philosophy that emphasizes the links (often the "continuity") between philosophy and science. For me, naturalism holds that the best way to address many philosophical problems is to approach them within our best current scientific picture of the world (chapter 10). Naturalism is sometimes taken to imply a claim about the ultimately physical nature of everything that exists. Naturalists are then thought to deny the existence of, for example, nonphysical souls. In this book I do not associate naturalism with any claims about what does and does not exist.

Normal science. In Kuhn's theory of science, this is the orderly form of science guided by a **paradigm**. Most science, for Kuhn, is normal science. A good normal scientist applies and does not usually question the fundamental ideas supplied by the paradigm. (chapter 5)

Objectivity. A term often used in a vague way to refer to beliefs or belief-forming procedures that avoid prejudice, caprice, and bias. A contrast is made with "subjective" beliefs or procedures, which bear the influence of a particular point of view. The term is also used to refer to a way in which things can be said to exist; something exists objectively if it exists independently of thought, language, or (again) a particular point of view. The two senses can be combined; avoiding bias and other subjective influences might be the best way to ensure that our beliefs represent things as they really are. (1.3, 8.4)

Occam's razor (also "Ockham's razor" or with other spellings). A principle associated with the medieval theologian William of Occam, which states that simpler theories should be preferred, when possible, to more complex ones. "Entities should not be multiplied beyond necessity." We should shave off everything that we don't need.

Operationalism (operationism). An empiricist view of science and scientific language developed by a physicist, Percy Bridgman, partly in response to Einstein's work in physics. According to operationalism, all good scientific language must either refer to observations or be definable in terms that refer only to observations. This view is similar to **logical positivism**, but is more a suggestion for how language should be used in science than a theory of meaning applied to all language. (2.3)

Paradigm. A term made famous by Kuhn's theory of science, but one he used in a number of ways. I distinguish two main senses. In the narrow sense, a paradigm is an impressive achievement that inspires and guides a tradition of further scientific work—a tradition of **normal science**. In the broad sense, a paradigm is a way of doing science that has grown up around a paradigm in the narrow sense. In the broad sense, a paradigm will typically include theoretical ideas about the world, methods, and subtle habits of mind and standards used to assess "good work" in the field. (chapter 5, chapter 6, section 7.7)

Parsimony. The "principle of parsimony" is another name for Occam's razor. Parsimony, more generally, is frugality.

Pessimistic induction from the history of science (pessimistic meta-induction). An argument against some forms of **scientific realism**. The argument holds that theories have changed so much in the history of science that we should not have much confidence in our current theories. In the past scientists have often been very confident that their theories were true, but (the argument says) they usually turned out to be wrong. We should therefore expect the same for our current theories. (10.6)

Phenomenalism. The view that when we seem to be talking and thinking about real physical objects, all we are really talking and thinking about are patterns in the flow of our sensations. The related word "phenomenon" is often used far more broadly than this strict meaning of "phenomenalism" would suggest; that word is used in philosophy with something like its everyday meaning, which is (roughly), something that happens. In science, the term "phenomenological law" is sometimes used to refer to a law of nature that does not provide a deep explanation but just describes a pattern or regularity.

Posterior probability. In **Bayesianism**, a probability of a hypothesis (h) given some piece of evidence (e). So it is a probability of the form $P(h|e)$. (chapter 12)

Pragmatism. A family of unorthodox empiricist philosophical views that emphasize the relation between thought and action. For pragmatists, the chief purpose of thought and language is practical problem-solving. The central figures in the "classical" stage of the pragmatist movement were Charles S. Peirce, William James, and John Dewey. Pragmatists reject the **correspondence** theory of **truth**. (12.2, 12.3, 14.1)

Prior probability. In **Bayesianism**, the initial or "unconditional" probability of a hypothesis (*h*) within an application of Bayes's theorem. So it is a probability of the form P(*h*). Bayes's theorem gives a formula for moving from the prior probability of a hypothesis to its **posterior probability**, the probability it has given some piece of evidence. (chapter 12)

Rationalism. In an older usage of this term, rationalism holds that some knowledge about the world can be gained through pure reasoning of a kind that does not depend on experience. Mathematics has been seen as an example. So in this sense, rationalism is opposed to **empiricism**. More recently, the term has been used for vaguer ideas that do not necessarily clash with empiricism. As a view about science, "rationalism" is often used for the idea that theory change is (or can be) guided by good reasoning and attention to evidence, as opposed to various kinds of bias or arbitrariness.

Realism. A wide variety of views can be described as "realist" in some sense, and debates about realism involve many different issues. Perhaps the most basic idea is this: a realist about something (or some kind of thing) thinks that it exists in a way that does not depend on our thoughts, language, or point of view. Questions about realism can be asked very broadly—perhaps about all facts, or about ordinary objects in the physical world. They can also be asked more narrowly, perhaps about numbers, moral facts, colors, or some other special category.

Relativism. The idea that the truth or justification of a claim, or the applicability of a standard or principle, depends on one's situation or point of view. Such a position can be asserted generally (about all truth or all standards) or specifically (about some particular domain, like morality or logic). The point of view might be that of an individual, a social group, the users of a particular language, or some other group. (5.6, 7.3, 8.4)

Research program. In Imre Lakatos's view of science, a sequence of scientific theories that all explore and develop the same basic theoretical ideas. Later theories in the sequence are developed in response to problems with the earlier ones. Some ideas in a research program—the "hard core"—are essential to the program and cannot be changed. Science typically involves ongoing competition between rival research programs in each field. (6.2)

Research tradition. Larry Laudan's research traditions are similar to Lakatos's **research programs**. There are differences, however, and Laudan's concept is probably more useful. For example, Laudan's research traditions include more than just theoretical ideas about the world; they include values and methods as well. For Laudan, the line between the fundamental ideas of a research tradition and the ever-changing details is not necessarily fixed. (6.3)

Scientific realism. A family of positions that assert some form of **realist** attitude toward the world as understood by science. Some views in this family hold that we should be confident that the world is accurately described by our well-tested scientific theories; other views hold just that science *aims* to give us descriptions of this kind. Scientific realists generally hold that theories at least aim to describe a world that exists in a mind-independent way. Scientific realism is opposed by metaphysical **constructivism**, **phenomenalism**, **operationalism**, and several other positions. (chapter 10)

Subjectivism (also personalism). An interpretation of the mathematics of probability theory, especially associated with **Bayesianism**. Subjectivists (at least of the strict kind) hold that probabilities are degrees of belief rather than measures of objective "chances" that exist in the world. Less strict versions of the view allow that there might be two kinds of probabilities, subjective ones and objective ones. (chapter 12)

Theory-ladenness of observation. A family of ideas that all claim, in some way, that observation cannot be an unbiased way to test rival theories (or larger units like paradigms) because observations, or their reports, are affected by the theoretical beliefs of the observer. (9.3)

Truth. In ordinary discussion, a true claim or sentence is one that describes how things really are; a false claim is one that misrepresents the world. Some, but not all, philosophical treatments of truth follow this familiar idea. **Correspondence** theories of truth agree with it. Other philosophers have tried to treat truth as depending only on what sort of evidence lies behind a claim or what sort of usefulness the claim has. Still others argue that we should not think of truth as a special relationship to the world (or our methods, or evidence) at all. Instead, "true" is just a linguistic tool used in discussion to express agreement and to make some other harmless moves. (chapter 13)

Verificationism. A theory of meaning associated with **logical positivism**. Verificationism is often summarized with the claim that the meaning of a sentence is its method of verification. "Verification" is a less appropriate word than "testing." Perhaps a better way to express the view is to say that to know the meaning of a sentence is the same thing as knowing how, in principle, to test it. The verificationist view of meaning is only applied to the parts of language that purport to describe the world (as opposed to expressing emotion, expressing commands, etc.) (chapter 2).

Websites

[1] http://blogs.plos.org/neuroanthropology/2010/12/01/anthropology -science-and-public-understanding/

[2] https://www.sciencemag.org/news/2016/10/gathering-hivaids-pioneers -raw-memories-mix-current-conflicts

[3] https://www.nytimes.com/2018/10/25/magazine/bruno-latour-post-truth -philosopher-science.html

[4] http://physics.nyu.edu/faculty/sokal/

[5] https://www.predictit.org

[6] https://www.nap.edu/read/771/chapter/2

References

Achinstein, Peter. 2001. *The Book of Evidence*. Oxford: Oxford University Press.

Alvarez, Luis W., Walter Alvarez, Frank Asaro, and Helen V. Michel. 1980. Extraterrestrial Cause for the Cretaceous-Tertiary Extinction. *Science* 208:1095–1108.

Appiah, Anthony. 1994. *Race, Culture, Identity: Misunderstood Connections*. Tanner Lectures on Human Values. San Diego: University of California Press.

Archibald, John. 2014. *One Plus One Equals One*. Oxford: Oxford University Press.

Armstrong, David M. 1983. *What Is a Law of Nature?* Cambridge: Cambridge University Press.

———. 1989. *Universals: An Opinionated Introduction*. Boulder, CO: Westview Press.

Ayer, Alfred J. 1936. *Language, Truth, and Logic*. London: V. Gollancz.

Baker, Alan. 2016. Simplicity. In *The Stanford Encyclopedia of Philosophy* (Winter ed.), edited by Edward N. Zalta. https://plato.stanford.edu/archives/win2016/entries/simplicity/.

Ball, Philip. 2013. *Curiosity: How Science Became Interested in Everything*. Chicago: University of Chicago Press.

Barnes, Barry, and David Bloor. 1982. Relativism, Rationalism, and the Sociology of Knowledge. In *Rationality and Relativism*, edited by Martin Hollis and Steven Lukes. Cambridge, MA: MIT Press.

Barnes, Barry, David Bloor, and John Henry. 1996. *Scientific Knowledge: A Sociological Analysis*. Chicago: University of Chicago Press.

Basu, Rima. 2018. The Wrongs of Racist Beliefs. *Philosophical Studies* 176:2497–2515.

Bayes, Thomas. 1763. An Essay towards Solving a Problem in the Doctrine of Chances. *Philosophical Transactions of the Royal Society of London* 53: 370–418.

Becker, Adam. 2018. *What Is Real? The Unfinished Quest for the Meaning of Quantum Physics*. London: John Murray.

Beebee, Helen. 2000. The Non-Governing Conception of Laws of Nature. *Philosophy and Phenomenological Research* 61:571–94.

Beebee, Helen, Christopher Hitchcock, and Peter C. Menzies, eds. 2009. *Oxford Handbook of Causation*. Oxford: Oxford University Press.

Berkeley, George. [1710 and 1713] 1988. *"Principles of Human Knowledge" and "Three Dialogues between Hylas and Philonous."* London and New York: Penguin Classics.

Biagioli, Mario, ed. 1999. *The Science Studies Reader*. New York: Routledge.

Blackburn, Simon. 2008. *The Oxford Dictionary of Philosophy*. Oxford: Oxford University Press.

———. 2018. *On Truth*. Oxford: Oxford University Press.

Bloor, David. 1976. *Knowledge and Social Imagery*. London: Routledge & Kegan Paul.

———. 1983. *Wittgenstein: A Social Theory of Knowledge*. London: Macmillan.

Boyd, Richard. 1984. The Current Status of Scientific Realism. In *Scientific Realism*, edited by Jarrett Leplin. Berkeley and Los Angeles: University of California Press.

———. 1991. Realism, Anti-Foundationalism, and the Enthusiasm for Natural Kinds. *Philosophical Studies* 61:127–48.

Boyer-Kassem, Thomas, Conor Mayo-Wilson, and Michael Weisberg, eds. 2017. *Scientific Collaboration and Collective Knowledge*. Oxford and New York: Oxford University Press.

Bridgman, Percy. [1927] 1991. The Operational Character of Scientific Concepts. In *The Philosophy of Science*, edited by Richard Boyd, Philip Gasper, and J. D. Trout. Cambridge, MA: MIT Press, 1991.

Bromberger, Sylvain. 1966. Why-Questions. In *Mind and Cosmos*, edited by Robert Colodny. Pittsburgh: University of Pittsburgh Press.

Campbell, Donald T. 1974. Evolutionary Epistemology. In *The Philosophy of Karl Popper*, edited by Paul Arthur Schilpp. La Salle, IL: Open Court.

Carnap, Rudolf. 1937. *The Logical Syntax of Language*. Translated by Amethe Smeaton. London: Routledge & Kegan Paul.

———. 1956. Empiricism, Semantics, and Ontology. In *Meaning and Necessity: A Study in Semantics and Modal Logic*. 2nd ed. Chicago: University of Chicago Press.

———. 1995. *An Introduction to the Philosophy of Science*. Edited by Martin Gardner. New York: Dover.

Carnap, Rudolf, Hans Hahn, and Otto Neurath. [1929] 1973. The Scientific Conception of the World: The Vienna Circle. In *Empiricism and Sociology*, edited by Marie Neurath and Robert S. Cohen. Dordrecht: Reidel.

Cartwright, Nancy. 1983. *How the Laws of Physics Lie*. Oxford: Oxford University Press.

Cartwright, Nancy, Jordi Cat, Lola Fleck, and Thomas E. Uebel. 1996. Otto Neurath: Philosophy between Science and Politics. *Philosophy* 71:632–34.

Chomsky, Noam. 1957. *Syntactic Structures*. The Hague: Mouton.

———. 1959. Review of B. F. Skinner's *Verbal Behavior*. *Language* 35:26–57.

Churchland, Paul M. 1988. Perceptual Plasticity and Theoretical Neutrality: A Reply to Jerry Fodor. *Philosophy of Science* 55:167–87.

Cohen, I. Bernard. 1985. *The Birth of a New Physics*. New York: W. W. Norton.

Cohen, Robert Sonné, Paul K. Feyerabend, and Marx W. Wartofsky, eds. 1976. *Essays in Memory of Imre Lakatos*. Boston Studies in the Philosophy of Science 39. Dordrecht: Reidel.

Cohen, Yael. 1987. Ravens and Relevance. *Erkenntnis* 26:153–79.

Copernicus, Nicolaus. [1543] 1992. *On the Revolutions*. Translated by Edward Rosen. Baltimore, MD: Johns Hopkins University Press.

Crick, Francis H. C. 1958. On Protein Synthesis. *Symposia of the Society for Experimental Biology* 12:138–63.

Darwin, Charles. [1859] 1964. *On the Origin of Species*. Edited by Ernst Mayr. Cambridge, MA: Harvard University Press.

Daston, Lorraine. 1995. Curiosity in Early Modern Science. *Word & Image* 11:391–404.

Daston, Lorraine, and Peter Galison. 2007. *Objectivity*. New York: Zone Books.

Dehaene, Stanislas. 2014. *Consciousness and the Brain: Deciphering How the Brain Codes Our Thoughts*. New York: Viking Adult.

Dennett, Daniel C. 1995. *Darwin's Dangerous Idea: Evolution and the Meanings of Life*. New York: Simon & Schuster.

Devitt, Michael. 1997. *Realism and Truth*. 2nd ed. Princeton, NJ: Princeton University Press.

Devlin, William J., and Alisa Bokulich. 2015. *Kuhn's Structure of Scientific Revolutions—50 Years On*. Cham: Springer International.

Dewey, John. 1929a. *Experience and Nature*. Rev. ed. La Salle, IL: Open Court.

———. 1929b. *The Quest for Certainty: A Study of the Relation of Knowledge and Action*. London: Allen & Unwin.

Doppelt, Gerald. 1978. Kuhn's Epistemological Relativism: An Interpretation and Defense. *Inquiry* 21:33–86.

Douglas, Heather. 2009. *Science, Policy, and the Value-Free Ideal*. Pittsburgh, PA: University of Pittsburgh Press.

Dowe, Philip. 1992. Process Causality and Asymmetry. *Erkenntnis* 37:179–96.

Downes, Stephen. 2020. *Models and Modeling in the Sciences: A Philosophical Introduction*. London: Routledge.

Dretske, Fred I. 1977. Laws of Nature. *Philosophy of Science* 44:248–68.

Duesberg, Peter H. 1987. Retroviruses as Carcinogens and Pathogens: Expectations and Reality. *Cancer Research* 47:1199–1220.

Dupré, John. 1993. *The Disorder of Things: Metaphysical Foundations of the Disunity of Science*. Cambridge, MA: Harvard University Press.

Earman, John. 1992. *Bayes or Bust? A Critical Examination of Bayesian Confirmation Theory*. Cambridge, MA: MIT Press.

Edmonds, David, and John Eidinow. 2001. *Wittgenstein's Poker: The Story of a Ten-Minute Argument between Two Great Philosophers*. New York: ECCO.

Einstein, Albert. [1916] 2015. *Relativity: The Special and the General Theory*. Translated by Robert Lawson. Princeton, NJ: Princeton University Press.

———. 1934. On the Method of Theoretical Physics. *Philosophy of Science* 1: 163–69.

Eldredge, Niles, and Stephen Jay Gould. 1972. Punctuated Equilibria: An Alternative to Phyletic Gradualism. In *Models in Paleobiology*, edited by Thomas J. M. Schopf. San Francisco: Freeman.

Elliott, Kevin C., and Ted Richards, eds. 2017. *Exploring Inductive Risk: Case Studies of Values in Science*. Oxford: Oxford University Press.

Evans, Alfred S. 1973. Pettenkofer Revisited: The Life and Contributions of Max von Pettenkofer (1818–1901). *Yale Journal of Biology and Medicine* 46:161–76.

Fanelli, Danielle. 2013. Why Growing Retractions Are (Mostly) a Good Sign. *PLOS Medicine* 10:e1001563.

Fang, Ferric C., R. Grant Steen, and Arturo Casadevall. 2012. Misconduct Accounts for the Majority of Retracted Scientific Publications. *Proceedings of the National Academy of the United States of America* 109:17028–33.

Faye, Jan. 2019. Copenhagen Interpretation of Quantum Mechanics. In *The Stanford Encyclopedia of Philosophy* (Winter ed.), edited by Edward N. Zalta. https://plato.stanford.edu/archives/win2019/entries/qm-copenhagen/.

Feder, Kenneth L. 1996. *Frauds, Myths, and Mysteries: Science and Pseudoscience in Archaeology*. Mountain View, CA: Mayfield.

Feigl, Herbert. 1943. Logical Empiricism. In *Twentieth Century Philosophy*, edited by Dagobert D. Runes. New York: Philosophical Library.

———. 1958. "The 'Mental' and the 'Physical.'" In *Concepts, Theories, and the Mind-Body Problem*, edited by Herbert Feigl, Michael Scriven, and Grover Maxwell, 370–497. Minnesota Studies in the Philosophy of Science 2. Minneapolis: University of Minnesota Press.

———. 1970. The "Orthodox" View of Theories: Remarks in Defense as Well As Critique. In *Theories and Methods of Physics and Psychology*, edited by Michael Radner and Stephen Winokur. Minnesota Studies in the Philosophy of Science 4. Minneapolis: University of Minnesota Press.

Feyerabend, Paul K. 1970. Consolations for the Specialist. In *Criticism and the Growth of Knowledge*, edited by Imre Lakatos and Alan Musgrave. Cambridge: Cambridge University Press.

———. 1975. *Against Method: Outline of an Anarchistic Theory of Knowledge*. Atlantic Highlands, NJ: Humanities Press.

———. 1978. *Science in a Free Society*. London: New Left Books.

———. 1981. *Philosophical Papers*. Cambridge: Cambridge University Press.

Feynman, Richard P., Robert B. Leighton, and Matthew Sands. 1963–65. *The Feynman Lectures on Physics*. Reading, MA: Addison-Wesley.

Fine, Arthur. 1984. The Natural Ontological Attitude. In *Scientific Realism*, edited by Jarrett Leplin. Berkeley and Los Angeles: University of California Press.

Fine, Cordelia. 2016. *Testosterone Rex: Unmaking the Myths of Our Gendered Minds*. New York: W. W. Norton.

de Finetti, Bruno [1931] 1989. Probabilism: A Critical Essay on the Theory of Probability and on the Value of Science. Translated by Maria Concetta Di Maio, Maria Carla Galavotti, and Richard C. Jeffrey. *Erkenntnis* 31:169–223.

Fodor, Jerry A. 1984. Observation Reconsidered. *Philosophy of Science* 51:23–43.

Forber, Patrick. 2011. Reconceiving Eliminative Inference. *Philosophy of Science* 78:185–208.

Forbes, Nancy, and Basil Mahon. 2014. *Faraday, Maxwell, and the Electromagnetic Field*. Amherst, NY: Prometheus Books.

Friedman, Michael. 1974. Explanation and Scientific Understanding. *Journal of Philosophy* 71:5–19.

———. 2001. *Dynamics of Reason*. Stanford Kant Lectures. Stanford, CA: CSLI Publications.

———. 2007. Kant—Naturphilosophie—Electromagnetism. In *Boston Studies in the Philosophy of Science, Hans Christian Ørsted, and the Romantic Legacy in Science: Ideas, Disciplines, Practices*, edited by Robert S. Cohen and Ole Knudsen. Dordrecht: Springer Netherlands.

Frigg, Roman. 2010. Models and Fiction. *Synthese* 172:251–68.

Frigg, Roman, and James Nguyen. 2016. Scientific Representation. In *The Stanford Encyclopedia of Philosophy* (Winter ed.), edited by Edward N. Zalta. https://plato.stanford.edu/archives/win2018/entries/scientific-representation/.

Futuyma, Douglas J. 1998. *Evolutionary Biology*. 3rd ed. Sunderland, MA: Sinauer Associates.

Galilei, Galileo. [1623] 1990. The Assayer. In *Discoveries and Opinions of Galileo*, translated by Stillman Drake. New York: Anchor Books.

———. [1632] 1967. *Dialogue Concerning the Two Chief World Systems, Ptolemaic & Copernican*. Berkeley and Los Angeles: University of California Press.

Galison, Peter. 1990. Aufbau/Bauhaus: Logical Positivism and Architectural Modernism. *Critical Inquiry* 16:709–52.

———. 1997. *Image and Logic: A Material Culture of Microphysics*. Chicago: University of Chicago Press.

Gannett, Lisa. 2003. Making Populations: Bounding Genes in Space and in Time. *Philosophy of Science* 70:989–1001.

Gendler, Tamar. 2011. On the Epistemic Costs of Implicit Bias. *Philosophical Studies* 156:33–63.

Chargaff, Erwin. 1951. Some Recent Studies on the Composition and Structure of Nucleic Acids. *Journal of Cellular and Comparative Physiology* 38: 41–59.

Giere, Ronald N. 1988. *Explaining Science: A Cognitive Approach*. Chicago: University of Chicago Press.

———. 2009. Why Scientific Models Should Not Be Regarded as Works of Fiction. In *Fictions in Science: Philosophical Essays on Modeling and Idealization*, edited by Mauricio Suárez. New York: Routledge.

Giere, Ronald N., and Alan W. Richardson, eds. 1997. *Origins of Logical Empiricism*. Minnesota Studies in the Philosophy of Science 16. Minneapolis: University of Minnesota Press.

Glymour, Clark N. 1980. *Theory and Evidence*. Princeton, NJ: Princeton University Press.

Godfrey-Smith, Peter. 2003. Goodman's Problem and Scientific Methodology. *Journal of Philosophy* 100:573–90.

———. 2009. Models and Fictions in Science. *Philosophical Studies* 143:101–16.

———. 2010. David Hull. *Biology & Philosophy* 25:749–53.

———. 2011. Induction, Samples, and Kinds. In *Carving Nature at Its Joints: Natural Kinds in Metaphysics and Science*, edited by Joseph Keim Campbell, Michael O'Rourke, and Matthew H. Slater. Cambridge, MA: MIT Press.

———. 2012. Darwinism and Cultural Change. *Philosophical Transactions of the Royal Society B* 367:2160–70.

———. 2014. Quine and Pragmatism. In *A Companion to W. V. O. Quine*, edited by Gilbert Harman and Ernest LePore. Malden, MA: Blackwell.

———. 2019. "Models, Fictions, and Conditionals." In *The Scientific Imagination: Philosophical and Psychological Perspectives*, edited by Arnon Levy and Peter Godfrey-Smith, 154–77. New York: Oxford University Press.

Goldman, Alvin I. 1999. *Knowledge in a Social World*. Oxford: Oxford University Press.

Good, I. J. 1967. The White Shoe Is a Red Herring. *British Journal for the Philosophy of Science* 17:322.

Goodman, Nelson. 1955. *Fact, Fiction and Forecast*. Cambridge, MA: Harvard University Press.

———. 1978. *Ways of Worldmaking*. Indianapolis, IN: Hackett.

———. 1996. Starmaking. In *Starmaking: Realism, Anti-Realism, and Irrealism*, edited by Peter McCormick. Cambridge, MA: MIT Press.

Gopnik, Alison, and Laura Schulz, eds. 2007. *Causal Learning: Psychology, Philosophy, and Computation*. Oxford: Oxford University Press.

Gould, Stephen Jay. 1977. Eternal Metaphors of Paleontology. In *Patterns of Evolution as Illustrated by the Fossil Record*, edited by Anthony Hallam. New York: Elsevier Scientific.

———. 1980. Is a New and General Theory of Evolution Emerging? *Paleobiology* 6:119–30.

———. 2002. *The Structure of Evolutionary Theory*. Cambridge, MA: Harvard University Press.

Greenwood, John D. 2015. *A Conceptual History of Psychology: Exploring the Tangled Web*. Cambridge: Cambridge University Press.

Gregory, Richard L. 1970. *The Intelligent Eye*. New York: McGraw-Hill.

Gross, Kevin, and Carl Bergstrom. 2019. Contest Models Highlight Inherent Inefficiencies of Scientific Funding Competitions. *PLOS Biology*. https://doi.org/10.1371/journal.pbio.3000065.

Gross, Paul R., and Norman Levitt. 1994. *Higher Superstition: The Academic Left and Its Quarrels with Science*. Baltimore MD: Johns Hopkins University Press.

Hacking, Ian. 1983. *Representing and Intervening: Introductory Topics in the Philosophy of Natural Science*. Cambridge: Cambridge University Press.

———. 1999. *The Social Construction of What?* Cambridge, MA: Harvard University Press.

Hájek, Alan, and James M. Joyce. 2008. Confirmation. In *The Routledge Companion to the Philosophy of Science*, edited by Stathis Psillos and Martin Curd. New York: Routledge.

Hall, Brian K. 2003. Evo-Devo: Evolutionary Developmental Mechanisms. *International Journal of Developmental Biology* 47:491–95.

Hanson, Norwood R. 1958. *Patterns of Discovery: An Inquiry into the Conceptual Foundations of Science*. Cambridge: Cambridge University Press.

Haraway, Donna. 1989. *Primate Visions: Gender, Race, and Nature in the World of Modern Science*. New York: Routledge.

Hardimon, Michael O. 2017. *Rethinking Race: The Case for Deflationary Realism*. Cambridge, MA: Harvard University Press.

Harding, Sandra G. 1986. *The Science Question in Feminism*. Ithaca, NY: Cornell University Press.

———. 1996. Rethinking Standpoint Epistemology: What Is "Strong Objectivity"? In *Feminism and Science*, edited by Evelyn Fox Keller and Helen E. Longino. Oxford: Oxford University Press.

Hardy, Godfrey H. 1940. *A Mathematician's Apology*. Cambridge: Cambridge University Press.

Harman, Gilbert H. 1965. Inference to the Best Explanation. *Philosophical Review* 74:88–95.

Harvey, David. 1989. *The Condition of Postmodernity: An Enquiry into the Origins of Cultural Change*. Oxford: Blackwell.

Heesen, Remco. 2018. Why the Reward Structure of Science Makes Reproducibility Problems Inevitable. *Journal of Philosophy* 115:661–74.

Hegel, Georg W. H. [1824] 1956. *The Philosophy of History*. New York: Dover.

Heidegger, Martin. [1927] 1962. *Being and Time*. Translated by John Macquarie and Edward Robinson. Oxford: Basil Blackwell.

Hempel, Carl G. 1958. The Theoretician's Dilemma. In *Concepts, Theories, and the Mind-Body Problem*, edited by Herbert Feigl, Michael Scriven, and Grover Maxwell. Minnesota Studies in the Philosophy of Science 2. Minneapolis: University of Minnesota Press.

———. 1965. Science and Human Values. In *Aspects of Scientific Explanation and Other Essays in the Philosophy of Science*. New York: Free Press.

———. 1966. *Philosophy of Natural Science*. Englewood Cliffs, NJ: Prentice-Hall.

Hempel, Carl G., and Paul Oppenheim. 1948. Studies in the Logic of Explanation. *Philosophy of Science* 15:135–75.

Henry, John. 1997. *The Scientific Revolution and the Origins of Modern Science*. New York: St. Martin's Press.

Hirsch, Jorge E. 2005. An Index to Quantify an Individual's Scientific Research Output. *Proceedings of the National Academy of Sciences* 102:16569–72.

Hitchcock, Christopher. 2019. Causal Models. In *The Stanford Encyclopedia of Philosophy* (Summer ed.), edited by Edward N. Zalta. https://plato.stanford.edu/archives/sum2019/entries/causal-models/.

Hobbes, Thomas. [1660] 1996. *Leviathan*. Edited by J. C. A. Gaskin. Oxford: Oxford University Press.

Hochman, Adam. 2017. Replacing Race: Interactive Constructionism about Racialized Groups. *Ergo* 4:61–92.

Hollis, Martin, and Steven Lukes, eds. 1982. *Rationality and Relativism*. Cambridge, MA: MIT Press.

Horgan, John. 1996. *The End of Science: Facing the Limits of Knowledge in the Twilight of the Scientific Age*. Reading, MA: Addison-Wesley.

Horwich, Paul. 1982. *Probability and Evidence*. Cambridge: Cambridge University Press.

———. 1990. *Truth*. Oxford: Clarendon Press.

———, ed. 1993. *World Changes: Thomas Kuhn and the Nature of Science*. Cambridge, MA: MIT Press.

Howson, Colin, and Peter Urbach. 2005. *Scientific Reasoning: The Bayesian Approach*. 3rd ed. Chicago: Open Court.

Hoyningen-Huene, Paul. 1993. *Reconstructing Scientific Revolutions: Thomas S. Kuhn's Philosophy of Science*. Chicago: University of Chicago Press.

Hrdy, Sarah Blaffer. 1999. *The Woman That Never Evolved*. Rev. ed. Cambridge, MA: Harvard University Press.

———. 2002. Empathy, Polyandry, and the Myth of the Coy Female. In *The Gender of Science*, edited by Janet A. Kourany. Upper Saddle River, NJ: Prentice Hall.

Hull, David L. 1988. *Science as a Process: An Evolutionary Account of the Social and Conceptual Development of Science.* Chicago: University of Chicago Press.

———. 1999. The Use and Abuse of Sir Karl Popper. *Biology and Philosophy* 14:481–504.

Hume, David. [1739] 1978. *A Treatise of Human Nature.* Edited by L. A. Selby-Bigge and P. H. Nidditch. Oxford: Oxford University Press.

———. [1740] 1978. An Abstract of *A Treatise of Human Nature.* In *A Treatise of Human Nature,* edited by L. A. Selby-Bigge and P. H. Nidditch. Oxford: Oxford University Press.

———. [1748] 1999. *An Enquiry Concerning Human Understanding.* Edited by Tom L. Beauchamp. Oxford: Oxford University Press.

Huntington, Samuel P. 1996. *The Clash of Civilizations and the Remaking of World Order.* New York: Simon & Schuster.

Jackson, Frank. 1975. Grue. *Journal of Philosophy* 72:113–31.

James, William. 1907. *Pragmatism: A New Name for Some Old Ways of Thinking.* Auckland: Floating Press.

Jeffrey, Richard C. 1956. Valuation and Acceptance of Scientific Hypotheses. *Philosophy of Science* 23:237–46.

———. 1965. *The Logic of Decision.* New York: McGraw-Hill.

Joyce, James M. 1998. A Nonpragmatic Vindication of Probabilism. *Philosophy of Science* 65:575–603.

Kant, Immanuel. [1781] 1998. *Critique of Pure Reason.* Translated by Paul Guyer and Allan W. Wood. Cambridge: Cambridge University Press.

———. [1786] 2002. *Metaphysical Foundations of Natural Science.* Translated by M. Friedman. In *Immanuel Kant: Theoretical Philosophy after 1781,* edited by Henry Allison, Peter Heath, and Gary Hatfield. Cambridge: Cambridge University Press.

Keller, Evelyn Fox. 1983. *A Feeling for the Organism: The Life and Work of Barbara McClintock.* San Francisco: W. H. Freeman.

———. 2002. A World of Difference. In *The Gender of Science,* edited by Janet A. Kourany. Upper Saddle River, NJ: Prentice Hall.

Keller, Evelyn Fox, and Helen E. Longino. 1996. *Feminism and Science.* Oxford: Oxford University Press.

Kelly, Kevin. 2004. Justification as Truth-Finding Efficiency: How Ockham's Razor Works. *Minds and Machines* 14:485–505.

Kimura, Motoo. 1983. *The Neutral Theory of Molecular Evolution.* Cambridge: Cambridge University Press.

Kincaid, Harold, John Dupré, and Alison Wylie, eds. 2007. *Value-Free Science? Ideas and Illusion.* Oxford: Oxford University Press.

Kitcher, Philip. 1981. Explanatory Unification. *Journal of Philosophy* 48: 507–31.

———. 1989. Explanatory Unification and the Causal Structure of the World. In *Scientific Explanation*, edited by Philip Kitcher and Wesley Salmon. Minnesota Studies in the Philosophy of Science 16. Minneapolis: University of Minnesota Press.

———. 1990. The Division of Cognitive Labor. *Journal of Philosophy* 87:5–22.

———. 1992. The Naturalists Return. *Philosophical Review* 101:53–114.

———. 1993. *The Advancement of Science*. Oxford: Oxford University Press.

———. 2001. *Science, Truth, and Democracy*. Oxford: Oxford University Press.

Klein, Jürgen, and Guido Giglioni. 2016. Francis Bacon. In *The Stanford Encyclopedia of Philosophy* (Winter ed.), edited by Edward N. Zalta. https://plato.stanford.edu/archives/win2016/entries/francis-bacon/.

Koertge, Noretta, ed. 1998. *A House Built on Sand: Exposing Postmodernist Myths about Science*. New York: Oxford University Press.

Koestler, Arthur. 1968. *The Sleepwalkers*. New York: Macmillan.

Kofman, Ava. 2018. Bruno Latour, the Post-Truth Philosopher, Mounts a Defense of Science. *New York Times*, October 25.

Kornblith, Hilary. 1993. *Inductive Inference and Its Natural Ground: An Essay in Naturalistic Epistemology*. Cambridge, MA: MIT Press.

———. 1994. *Naturalizing Epistemology*. 2nd ed. Cambridge, MA: MIT Press.

Kourany, Janet. 2002. *The Gender of Science*. Upper Saddle River, NJ: Prentice Hall.

Kuhn, Thomas S. 1957. *The Copernican Revolution: Planetary Astronomy in the Development of Western Thought*. Cambridge, MA: Harvard University Press.

———. [1962] 1996. *The Structure of Scientific Revolutions*. 3rd ed. Chicago: University of Chicago Press.

———. 1970a. "Postscript" to *The Structure of Scientific Revolutions*. University of Chicago Press.

———. 1970b. Reflections on My Critics. In *Criticism and the Growth of Knowledge*, edited by Imre Lakatos and Alan Musgrave. Cambridge: Cambridge University Press.

———. 1977a. Concepts of Cause in the Development of Physics. In *The Essential Tension: Selected Studies in Scientific Tradition and Change*. Chicago: University of Chicago Press.

———. 1977b. *The Essential Tension: Selected Studies in Scientific Tradition and Change*. Chicago: University of Chicago Press.

———. 1977c. Objectivity, Value Judgment, and Theory Choice. In *The Essential*

Tension: Selected Studies in Scientific Tradition and Change. Chicago: University of Chicago Press.

———. 1978. *Black-Body Theory and the Quantum Discontinuity, 1894–1912.* Oxford: Oxford University Press.

———. 2000. *The Road since Structure: Philosophical Essays, 1970–1993, with an Autobiographical Interview.* Edited by James Conant and John Haugeland. Chicago: University of Chicago Press.

Lakatos, Imre. 1970. Falsification and the Methodology of Scientific Research Programmes. In *Criticism and the Growth of Knowledge,* edited by Imre Lakatos and Alan Musgrave. Cambridge: Cambridge University Press.

———. 1971. History of Science and Its Rational Reconstructions. In *PSA 1970,* edited by Roger C. Buck and Robert S. Cohen. Dordrecht: Reidel.

———. 1976. *Proofs and Refutations: The Logic of Mathematical Discovery.* Cambridge: Cambridge University Press.

Lakatos, Imre, and Alan Musgrave, eds. 1970. *Criticism and the Growth of Knowledge.* Cambridge: Cambridge University Press.

Lanctot, Richard, Kim Scribner, Bart Kempenaers, and Patrick Weatherhead. 1997. Lekking without a Paradox in the Buff-Breasted Sandpiper. *American Naturalist* 149:1051–70.

Latour, Bruno. 1987. *Science in Action: How to Follow Scientists and Engineers through Society.* Cambridge, MA: Harvard University Press.

———. 1988. *The Pasteurization of France.* Cambridge, MA: Harvard University Press.

———. 1993. *We Have Never Been Modern.* Cambridge, MA: Harvard University Press.

Latour, Bruno, and Steve Woolgar. 1979. *Laboratory Life: The Construction of Scientific Facts.* Beverly Hills, CA: Sage.

Laudan, Larry. 1977. *Progress and Its Problems: Toward a Theory of Scientific Growth.* Berkeley and Los Angeles: University of California Press.

———. 1981. A Confutation of Convergent Realism. *Philosophy of Science* 48: 19–48.

———. 1987. Progress or Rationality? The Prospects for Normative Naturalism. *American Philosophical Quarterly* 24:19–31.

Leplin, Jarrett, ed. 1984. *Scientific Realism.* Berkeley and Los Angeles: University of California Press.

Levy, Arnon. 2015. Modeling without Models. *Philosophical Studies* 172:781–98.

Levy, Arnon, and Peter Godfrey-Smith, eds. 2019. *The Scientific Imagination: Philosophical and Psychological Perspectives.* New York: Oxford University Press.

Lewis, David. 1986a. Causation and Explanation. In *Philosophical Papers*, vol. 2. Oxford: Oxford University Press.

———. 1986b. Introduction to *Philosophical Papers*, vol. 2. Oxford: Oxford University Press.

Lewontin, Richard C. 1972. The Apportionment of Human Diversity. In *Evolutionary Biology*, edited by Theodosius Dobzhansky, Max K. Hecht, and William C. Steere. New York: Springer.

Lingua Franca Editors. 2000. *The Sokal Hoax: The Sham That Shook the Academy*. Lincoln: University of Nebraska Press.

Lipton, Peter. 1991. *Inference to the Best Explanation*. London: Routledge.

Lloyd, Elisabeth A. 1993. Pre-Theoretical Assumptions in Evolutionary Explanations of Female Sexuality. *Philosophical Studies* 69:139–53.

———. 1997. Feyerabend, Mill, and Pluralism. *Philosophy of Science* 64:S396–S408.

———. 2005. *The Case of the Female Orgasm: Bias in the Science of Evolution*. Cambridge, MA: Harvard University Press.

Lloyd, Genevieve. 1984. *The Man of Reason: "Male" and "Female" in Western Philosophy*. Minneapolis: University of Minnesota Press.

Locke, John. [1690] 1975. *An Essay Concerning Human Understanding*. Edited by Peter H. Nidditch. Oxford: Clarendon Press.

Longino, Helen E. 1990. *Science as Social Knowledge: Values and Objectivity in Scientific Inquiry*. Princeton, NJ: Princeton University Press.

Lyotard, Jean-Francois. [1979] 1984. *The Postmodern Condition: A Report on Knowledge*. Translated by Geoff Bennington and Brian Massumi. Minneapolis: University of Minnesota Press.

Mach, Ernst. 1897. *Contributions to the Analysis of the Sensations*. Translated by C. M. Williams. Chicago: Open Court.

Machamer, Peter, Lindley Darden, and Carl F. Craver. 2000. Thinking about Mechanisms. *Philosophy of Science* 67:1–25.

MacKenzie, Donald A. 1981. *Statistics in Britain, 1865–1930: The Social Construction of Scientific Knowledge*. Edinburgh: Edinburgh University Press.

Mackie, John L. 1980. *The Cement of the Universe: A Study of Causation*. 2nd ed. Oxford: Oxford University Press.

Marshall, Lisa, Halla Helgadóttir, Matthias Mölle, and Jan Born. 2006. Boosting Slow Oscillations during Sleep Potentiates Memory. *Nature* 444:610–13.

Maudlin, Tim. 2019. *Philosophy of Physics: Quantum Theory*. Princeton, NJ: Princeton University Press.

Maxwell, Grover. 1962. The Ontological Status of Theoretical Entities. In *Scientific Explanation, Space, and Time*, edited by Herbert Feigl and Grover

Maxwell. Minnesota Studies in the Philosophy of Science 3. Minneapolis: University of Minnesota Press.

Maynard Smith, John, and George R. Price. 1973. The Logic of Animal Conflict. *Nature* 246:15–18.

Mayo, Deborah G. 1996. *Error and the Growth of Experimental Knowledge.* Chicago: University of Chicago Press.

McMullin, Ernan. 1984. A Case for Scientific Realism. In *Scientific Realism,* edited by Jarrett Leplin. Berkeley and Los Angeles: University of California Press.

Merton, Robert K. [1957] 1973. Priorities in Scientific Discovery. In *The Sociology of Science: Theoretical and Empirical Investigations,* edited by Norman Storer. Chicago: University of Chicago Press.

———. 1973. *The Sociology of Science: Theoretical and Empirical Investigations.* Edited by Norman Storer. Chicago: University of Chicago Press.

Mill, John Stuart. [1859] 1978. *On Liberty.* Edited by Elizabeth Rapaport. Indianapolis: Hackett.

———. 1865. *An Examination of Sir William Hamilton's Philosophy and of the Principal Philosophical Questions Discussed in His Writings.* Boston: W. V. Spencer.

Miller, George A., Eugene Galanter, and Karl H. Pribram. 1960. *Plans and the Structure of Behavior.* New York: Henry Holt.

Millikan, Ruth G. 2017. *Beyond Concepts: Unicepts, Language, and Natural Information.* Oxford: Oxford University Press.

Miner, Valerie, and Helen E. Longino, eds. 1987. *Competition: A Feminist Taboo?* New York: Feminist Press at CUNY.

Misak, Cheryl. 2020. *Frank Ramsey: A Sheer Excess of Powers.* Oxford: Oxford University Press.

Mitchell, Sandra D. 2000. Dimensions of Scientific Law. *Philosophy of Science* 67: 242–65.

Motterlini, Matteo, ed. 1999. *For and Against Method.* Chicago: University of Chicago Press.

Moss, Sarah. 2018. Moral Encroachment. In *Proceedings of the Aristotelian Society,* vol. 118, part 2. DOI: 10.1093/arisoc/aoy007.

Musgrave, Alan. 1976. Method or Madness. In *Essays in Memory of Imre Lakatos,* edited by Robert Sonné Cohen, Paul K. Feyerabend, and Marx W. Wartofsky. Dordrecht: Reidel.

Musgrave, Alan, and Charles Pigden. 2016. Imre Lakatos. In *The Stanford Encyclopedia of Philosophy* (Winter ed.), edited by Edward N. Zalta. https://plato.stanford.edu/archives/win2016/entries/lakatos/.

Nattrass, Nicoli. 2008. AIDS and the Scientific Governance of Medicine in Post-Apartheid South Africa. *African Affairs* 107:157–76.

Nedergaard, Jan, Tore Benson, and Barbara Cannon. 2007. Unexpected Evidence for Active Brown Adipose Tissue in Adult Humans. *American Journal of Physiology* 293:444–52.

Neurath, Otto. 1921. *Anti-Spengler*. Munich: Callwey.

Newell, Allen, John C. Shaw, and Herbert A. Simon. 1958. Elements of a Theory of Human Problem Solving. *Psychological Review* 65:151–66.

Newton, Isaac. [1687] 1999. *The Principia: Mathematical Principles of Natural Philosophy*. Translated by I. Bernard Cohen and Anne M. Whitman. Berkeley and Los Angeles: University of California Press.

Newton-Smith, William H. 1981. *The Rationality of Science*. Boston: Routledge & Kegan Paul.

Norton, John D. 2003. A Material Theory of Induction. *Philosophy of Science* 70:647–70.

Okasha, Samir. 2011. Experiment, Observation and the Confirmation of Laws. *Analysis* 71:222–32.

O'Keefe, John, and Lynn Nadel. 1978. *The Hippocampus as a Cognitive Map*. Oxford: Clarendon Press.

Okruhlik, Kathleen. 1994. Gender and the Biological Sciences. *Canadian Journal of Philosophy* 24:21–42.

Oppenheimer, Michael, Naomi Oreskes, Dale Jamieson, Keynyn Brysse, Jessica O'Reilly, Matthew Shindell, and Milena Wazeck. 2019. *Discerning Experts: The Practices of Scientific Assessment for Environmental Policy*. Chicago: University of Chicago Press.

Pearl, Judea. 2000. *Causality: Models, Reasoning, and Inference*. Cambridge: Cambridge University Press.

Pearl, Judea, and Dana Mackenzie. 2018. *The Book of Why: The New Science of Cause and Effect*. New York: Basic Books.

Peirce, Charles S. 1877. The Fixation of Belief. *Popular Science Monthly* 12:1–15.

———. 1878. How to Make Our Ideas Clear. *Popular Science Monthly* 12: 286–302.

———. [1898] 1993. *Reasoning and the Logic of Things: The Cambridge Conferences Lectures of 1898*, edited by Kenneth Laine Ketner. Cambridge, MA: Harvard University Press.

Pigliucci, Massimo, and Jonathan Kaplan. 2003. On the Concept of Biological Race and Its Applicability to Humans. *Philosophy of Science* 70:1161–72.

Pinker, Steven. 2011. *The Better Angels of Our Nature*. New York: Viking Press.

Platt, John R. 1964. Strong Inference. *Science* 146:347–53.

Popper, Karl R. 1935. *Logik der Forschung: Zur Erkenntnistheorie der modernen Naturwissenschaft*. Vienna: J. Springer.

———. 1959. *The Logic of Scientific Discovery*. New York: Basic Books.

———. 1963. *Conjectures and Refutations: The Growth of Scientific Knowledge*. London: Routledge & Kegan Paul.

———. 1970. Normal Science and Its Dangers. In *Criticism and the Growth of Knowledge*, edited by Imre Lakatos and Alan Musgrave. Cambridge: Cambridge University Press.

———. 1976. The Myth of the Framework. In *Rational Changes in Science: Essays on Scientific Reasoning*, edited by Joseph C. Pitt and Marcello Pera. Dordrecht: Reidel.

Porter, Roy. 1998. *The Greatest Benefit to Mankind: A Medical History of Humanity*. New York: W. W. Norton.

Potochnik, Angela. 2017. *Idealization and the Aims of Science*. Chicago: University of Chicago Press.

Preston, John, Gonzalo Munévar, and David Lamb, eds. 2000. *The Worst Enemy of Science? Essays in Memory of Paul Feyerabend*. Oxford: Oxford University Press.

Price, Huw. 2003. Truth as Convenient Friction. *Journal of Philosophy C* 100: 167–90.

———. 2005. Causal Perspectivism. In *Causation, Physics, and the Constitution of Reality: Russell's Republic Revisited*, edited by Huw Price and Richard Corry. Oxford: Oxford University Press.

Proietti, Massimiliano, Alexander Pickson, Francesco Graffitti, Peter Barrow, Dmytro Kundys, Cyril Branciard, Martin Ringbauer, and Alessandro Fedrizzi. 2019. Experimental Test of Local Observer Independence. *Science Advances* 5:eaaw9832.

Provine, William B. 1971. *The Origins of Theoretical Population Genetics*. Chicago: University of Chicago Press.

Psillos, Stathis. 1999. *Scientific Realism: How Science Tracks Truth*. New York: Routledge.

Putnam, Hilary. 1975a. *Mathematics, Matter and Method*. Cambridge: Cambridge University Press.

———. 1975b. *Mind, Language, and Reality*. Cambridge: Cambridge University Press.

Quine, Willard V. 1951a. Two Dogmas of Empiricism. *Philosophical Review* 60:20–43. Reprinted with slight differences in *From a Logical Point of View*. Cambridge. MA: Harvard University Press, 1953.

———. 1951b. On Carnap's Views on Ontology. *Philosophical Studies* 2:65–72.

———. 1969. Epistemology Naturalized. In *Ontological Relativity and Other Essays*. New York: Columbia University Press.

———. 1990. *Pursuit of Truth*. Cambridge, MA: Harvard University Press.

Railton, Peter. 1981. Probability, Explanation, and Information. *Synthese* 48: 231–56.

Ramsey, Frank Plumpton. [1926] 1931. Truth and Probability. In *The Foundations of Mathematics and Other Logical Essays*, edited by R. B. Braithwaite. London: Kegan, Paul, Trench, Trubner.

Reichenbach, Hans. 1938. *Experience and Prediction: An Analysis of the Foundations and the Structure of Knowledge*. Chicago: University of Chicago Press.

———. 1951. *The Rise of Scientific Philosophy*. Berkeley and Los Angeles: University of California Press.

Rensberger, Boyce. 1977. Fraud in Research Is a Rising Problem in Science. *New York Times*, January 23.

Resnik, Michael D. 1987. *Choices: An Introduction to Decision Theory*. Minneapolis: University of Minnesota Press.

Rhodes, Richard. 1987. *The Making of the Atomic Bomb*. New York: Simon & Schuster.

Rinard, Susanna. 2014. A New Bayesian Solution to the Paradox of the Ravens. *Philosophy of Science* 81:81–100.

———. 2019. Equal Treatment for Belief. *Philosophical Studies* 176:1923–50.

Rorty, Richard. 1982. *Consequences of Pragmatism*. Minneapolis: University of Minnesota Press.

Ross, Lauren N., and James F. Woodward. 2016. Koch's Postulates: An Interventionist Perspective. *Studies in History and Philosophy of Science C: Studies in History and Philosophy of Biological and Biomedical Sciences* 59:35–46.

Rudner, Richard. 1953. The Scientist qua Scientist Makes Value Judgments. *Philosophy of Science* 20:1–6.

Sagan, Lynn. 1967. On the Origin of Mitosing Cells. *Journal of Theoretical Biology* 14:225–74.

Salmon, Wesley C. 1981. Rational Prediction. *British Journal for the Philosophy of Science* 32:115–25.

———. 1984. *Scientific Explanation and the Causal Structure of the World*. Princeton, NJ: Princeton University Press.

———. 1989. *Four Decades of Scientific Explanation*. Minneapolis: University of Minnesota Press.

———. 1990. Rationality and Objectivity in Science, or, Tom Kuhn meets Tom

Bayes. In *Scientific Theories*, edited by C. Wade Savage. Minneapolis: University of Minnesota Press.

———. 1998. Scientific Explanation: Causation and Unification. In *Causality and Explanation*. Oxford: Oxford University Press.

Schilpp, Paul Arthur, ed. 1974. *The Philosophy of Karl Popper*. La Salle, IL: Open Court.

Schlick, Moritz. 1932–33. Positivism and Realism. *Erkenntnis* 3:1–31 (in German). Translated by Peter Heath and reprinted in *The Philosophy of Science*, edited by Richard Boyd, Philip Gasper, and J. D. Trout. Cambridge, MA: MIT Press, 1991.

Schrödinger, Erwin. 1958. *Mind and Matter*. Cambridge: Cambridge University Press.

Schulz, Kathryn. 2010. *Being Wrong: Adventures in the Margin of Error*. London: Portobello.

Sellars, Wilfrid. 1962. Philosophy and the Scientific Image of Man. In *Frontiers of Science and Philosophy*, edited by Robert Colodny. Pittsburgh: University of Pittsburgh Press.

Shapin, Steven. 1982. History of Science and Its Sociological Reconstructions. *History of Science* 20:157–211.

———. 1994. *A Social History of Truth: Civility and Science in Seventeenth-Century England*. Chicago: University of Chicago Press.

———. 1996. *The Scientific Revolution*. Chicago: University of Chicago Press.

Shapin, Steven, and Simon Schaffer. 1985. *Leviathan and the Air-Pump: Hobbes, Boyle, and the Experimental Life*. Princeton, NJ: Princeton University Press.

Shearmur, Geremy, and Geoff Stokes, eds. 2016. *The Cambridge Companion to Popper*. Cambridge and New York: Cambridge University Press.

Sheehan, Thomas. 1998. Heidegger, Martin (1889–1976). In *Routledge Encyclopedia of Philosophy*, vol. 2, edited by Edward Craig. New York: Routledge.

Shimony, Abner. 1996. Some Historical and Philosophical Reflections on Science and Enlightenment. *Philosophy of Science* 64:1–14.

Skinner, B. F. 1938. *The Behavior of Organisms: An Experimental Analysis*. New York: Appleton-Century.

Skyrms, Brian. 2000. *Choice and Chance: An Introduction to Inductive Logic*. 3rd ed. Belmont, CA: Wadsworth/Thomson Learning.

Smaldino, Paul E., and Richard McElreath. 2016. The Natural Selection of Bad Science. *Royal Society Open Science* 3:160384.

Smart, J. J. C. 1963. *Philosophy and Scientific Realism*. New York: Humanities Press.

Smith, Adam. [1776] 1976. *An Inquiry into the Nature and Causes of the Wealth of Nations*. Edited by R. H. Campbell and A. S. Skinner. Oxford: Oxford University Press.

Sobel, Dava. 1999. *Galileo's Daughter: A Historical Memoir of Science, Faith, and Love*. New York: Walker.

Sober, Elliott. 1994. Let's Razor Ockham's Razor. In *From a Biological Point of View: Essays in Evolutionary Philosophy*. Cambridge: Cambridge University Press.

———. 2015. *Ockham's Razors: A User's Manual*. Cambridge: Cambridge University Press.

Sokal, Alan. 1996a. A Physicist Experiments with Cultural Studies. *Lingua Franca* (May–June): 62–64.

———. 1996b. Transgressing the Boundaries: Toward a Transformative Hermeneutics of Quantum Gravity. *Social Text* 14:217–52.

Solomon, Miriam. 2001. *Social Empiricism*. Cambridge, MA: MIT Press.

Spencer, Quayshawn. 2018a. Racial Realism I: Are Biological Races Real? *Philosophy Compass* 13:e12468. https://doi.org/10.1111/phc3.12468.

———. 2018b. Racial Realism II: Are Folk Races Real? *Philosophy Compass* 13: e12467. https://doi.org/10.1111/phc3.12467.

Spirtes, Peter, Clark N. Glymour, and Richard Scheines. 1993. *Causation, Prediction, and Search*. New York: Springer-Verlag.

Stalker, Douglas Frank. 1994. *Grue! The New Riddle of Induction*. Chicago: Open Court.

Stanford, Kyle. 2006. *Exceeding Our Grasp: Science, History, and the Problem of Unconceived Alternatives*. Oxford: Oxford University Press.

Steen, R. Grant. 2011. Retractions in the Scientific Literature: Is the Incidence of Research Fraud Increasing? *Journal of Medical Ethics* 37:249–53.

Steen, R. Grant, Arturo Casadevall, and Ferric C. Fang. 2016. Why Has the Number of Scientific Retractions Increased? In *Methodological Issues and Strategies in Clinical Research*, edited by Alan E. Kazdin. Washington, DC: American Psychological Association.

Sterrett, Susan G. 2002. Physical Models and Fundamental Laws: Using One Piece of the World to Tell about Another. *Mind & Society* 3:51–66.

Strevens, Michael. 2003. The Role of the Priority Rule in Science. *Journal of Philosophy* 100:55–79.

———. 2004. The Causal and Unification Accounts of Explanation Unified—Causally. *Noûs* 38:154–79.

———. 2008. *Depth: An Account of Scientific Explanation*. Cambridge, MA: Harvard University Press.

Suárez, Mauricio. 2009. Scientific Fictions as Rules of Inference. In *Fictions in Science: Philosophical Essays on Modeling and Idealization.* London: Routledge.

Suppes, Patrick. 1984. *Probabilistic Metaphysics.* Oxford: Blackwell.

Thomson-Jones, Martin. 2010. Missing Systems and the Face Value Practice. *Synthese* 172:283–99.

Tolman, Edward C. 1948. Cognitive Maps in Rats and Men. *Psychological Review* 55:189–208.

Toulmin, Stephen. 1972. *Human Understanding.* Princeton, NJ: Princeton University Press.

Toulmin, Stephen Edelston, and June Goodfield. 1962. *The Fabric of the Heavens: The Development of Astronomy and Dynamics.* New York: Harper.

———. 1982. *The Architecture of Matter.* Chicago: University of Chicago Press.

van Fraassen, Bas C. 1980. *The Scientific Image.* Oxford: Oxford University Press.

Wason, Peter Cathcart, and Philip N. Johnson-Laird. 1972. *Psychology of Reasoning: Structure and Content.* Cambridge, MA: Harvard University Press.

Weisberg, Michael. 2007. Who Is a Modeler? *British Journal for the Philosophy of Science* 58:207–33.

———. 2013. *Simulation and Similarity: Using Models to Understand the World.* New York: Oxford University Press.

Westfall, Richard S. 1980. *Never at Rest: A Biography of Isaac Newton.* Cambridge: Cambridge University Press.

———. 1993. *The Life of Isaac Newton.* Cambridge: Cambridge University Press.

Whitcomb, Dennis, and Alvin Goldman, eds. 2011. *Social Epistemology: Essential Readings.* Oxford: Oxford University Press.

Wigner, Eugene P. 1960. The Unreasonable Effectiveness of Mathematics in the Natural Sciences. *Communications on Pure and Applied Mathematics* 13:1–14.

———. [1961] 1995. Remarks on the Mind-Body Question. In *Philosophical Reflections and Syntheses: The Collected Works of Eugene Paul Wigner, edited by Jagdish Mehra.* Vol. 6 of *Part B: Historical, Philosophical, and Socio-Political Papers.* Berlin: Springer.

Williams, Michael. 1998. Feyerabend, Paul Karl (1924–94). In *Routledge Encyclopedia of Philosophy*, vol. 2, edited by Edward Craig, 640–42. New York: Routledge.

Winther, Rasmus Grønfeldt. 2009. Prediction in Selectionist Evolutionary Theory. *Philosophy of Science* 76:889–901.

Wittgenstein, Ludwig. [1922] 1988. *Tractatus Logico-Philosophicus*. Translated by David Pears and Brian McGuinness. London: Routledge & Kegan Paul.

———. 1953. *Philosophical Investigations*. Translated by G. E. M. Anscombe. New York: Macmillan.

Wolfe, Tom. 1998. *A Man in Full: A Novel*. New York: Farrar, Straus & Giroux.

Woodward, James. 2003. *Making Things Happen: A Theory of Causal Explanation*. New York: Oxford University Press.

Woolgar, Steven. 1988. *Science: The Very Idea*. London: Ellis Horwood.

Worrall, John. 1989. Structural Realism: The Best of Both Worlds? *Dialectica* 43:99–124.

Zollman, Kevin J. S. 2010. The Epistemic Benefit of Transient Diversity. *Erkenntnis* 72:17–35.

Index

A page number in italics refers to a figure.

Made in United States
North Haven, CT
24 August 2023

40700153R10225